Wireless Networks

Series Editor

Xuemin Sherman Shen, University of Waterloo, Waterloo, Canada

The purpose of Springer's Wireless Networks book series is to establish the state of the art and set the course for future research and development in wireless communication networks. The scope of this series includes not only all aspects of wireless networks (including cellular networks, WiFi, sensor networks, and vehicular networks), but related areas such as cloud computing and big data. The series serves as a central source of references for wireless networks research and development. It aims to publish thorough and cohesive overviews on specific topics in wireless networks, as well as works that are larger in scope than survey articles and that contain more detailed background information. The series also provides coverage of advanced and timely topics worthy of monographs, contributed volumes, textbooks and handbooks.

Yong Zhou • Wenzhi Fang • Yuanming Shi •
Khaled B. Letaief

Federated Edge Learning

Algorithms, Architectures and Trustworthiness

Springer

Yong Zhou
Shanghai, China

Wenzhi Fang
Shanghai, China

Yuanming Shi
Shanghai, China

Khaled B. Letaief
Hong Kong, Hong Kong

ISSN 2366-1186 ISSN 2366-1445 (electronic)
Wireless Networks
ISBN 978-3-031-96648-4 ISBN 978-3-031-96649-1 (eBook)
https://doi.org/10.1007/978-3-031-96649-1

This Springer imprint is published by the registered company Springer Nature Switzerland AG
The registered company address is: Gewerbestrasse 11, 6330 Cham, Switzerland

Preface

The thriving of various intelligent applications (e.g., autonomous driving, smart city) is an essential driving force for the paradigm shift from "connected things" to "connected intelligence" in 6G. To support scalable and trustworthy "connected intelligence," federated edge learning (FEEL) is emerging as a disruptive collaborative machine learning framework, given its capability to achieve privacy-preserving model training by only exchanging local model updates instead of any private raw data. Despite the great promise, supporting FEEL over wireless networks still confronts the challenges of severe communication bottlenecks due to the periodic exchange of high-dimensional model updates, slow convergence due to the statistical and system heterogeneity, as well as the trustworthiness issue due to inference and Byzantine attack. To tackle these issues, various effective communication and learning schemes need to be developed. This book intends to present various effective schemes from the perspectives of algorithms, architectures, as well as privacy and security to enable scalable and trustworthy FEEL.

From the algorithmic perspective, we elaborate various federated optimization algorithms in Chaps. 2–4, including zeroth-order, first-order, and second-order methods, with specific emphasis on presenting provable convergence analysis to illustrate the impact of learning and wireless communication parameters. Specifically, we first present a federated first-order optimization algorithm to update the global model for FEEL in Chap. 2, where over-the-air computation (AirComp) is adopted to support fast local model aggregation, and device selection and beamforming design are jointly optimized to enhance the learning performance. Subsequently, we present a federated second-order optimization algorithm in Chap. 3 to reduce the number of communication rounds for convergence, at the cost of increasing the local computation burden in each communication round. Rigorous theoretical analysis is provided to show that the proposed second-order algorithm for FEEL under data heterogeneity keeps a linear-quadratic convergence rate. Although effective, both the federated first-order and second-order algorithms cannot be applied in scenarios where the gradient information is not available. As a result, we present a federated zeroth-order optimization algorithm in Chap. 4, where the mini-batch-type stochastic gradient estimator is utilized. The convergence of the proposed federated zeroth-order optimization algorithm is established for both the error-free and fading channels. The computation complexity and communication overhead of the federated zeroth/first/second-order algorithms over wireless networks are also elaborated.

From the network architecture perspective, we elaborate how the critical challenges of FEEL can be addressed by exploiting different architectures and designing effective communication schemes in Chaps. 5–7. The communication straggler issue of FEEL can be mitigated by reconfiguring the propagation environment by utilizing reconfigurable intelligent surface (RIS) and unmanned aerial vehicle (UAV). Specifically, we propose an RIS-assisted FEEL in Chap. 5 to relieve the communication bottleneck and characterize the corresponding convergence behavior, followed by developing a graph neural network-based learning algorithm to directly optimize the RIS phase shifts and the transceiver. Subsequently, we leverage UAV's high altitude and mobility to mitigate the communication bottleneck of FEEL in Chap. 6. By illustrating the impact of device scheduling on the convergence performance, we jointly optimize the UAV trajectory and device scheduling to minimize the training latency. Moreover, the multi-cell architecture in Chap. 7 presents a feasible solution for collaborative FEEL among multiple cells.

Finally, we discuss the challenges of FEEL from the privacy and security perspective in Chaps. 8–9, followed by presenting effective communication schemes that can achieve differentially private model aggregation and Byzantine-resilient model aggregation to achieve trustworthy FEEL. Specifically, we propose an RIS-assisted FEEL approach to balance the tradeoff between differential privacy and learning accuracy in Chap. 8. We evaluate the convergence performance of the proposed algorithm, followed by developing a two-step alternating minimization algorithm to enhance the learning performance while adhering to the privacy constraint. Additionally, we present a decentralized blockchain-based FEEL architecture in Chap. 9, where a secure global aggregation algorithm and a Byzantine fault tolerance consensus protocol are developed to resist malicious attacks.

In summary, this book provides a comprehensive discussion on FEEL, spanning from communication-efficient federated optimization algorithms, promising communication techniques and network architectures, learning-aware resource allocation algorithms, to trustworthy model aggregation schemes. We hope that this book will enlighten more designs of algorithms, architectures, techniques for FEEL and attract more researchers into this promising field.

Shanghai, China Yong Zhou
Shanghai, China Wenzhi Fang
Shanghai, China Yuanming Shi
Hong Kong, China Khaled B. Letaief

Declarations

Competing Interests The authors have no competing interests to declare that are relevant to the content of this manuscript.

Contents

Acronyms

AI	Artificial Intelligence
AirComp	Over-the-Air Computation
AOA	Angles of Arrival
AOD	Angles of Departure
AWGN	Additive White Gaussian Noise
B-FEEL	Blockchain-based Federated Edge Learning
BFGS	Broyden-Fletcher-Goldfarb-Shanno
BS	Base Station
CNNs	Convolutional Neural Networks
CSI	Channel State Information
DC	Difference-of-Convex Functions
DCA	DC Algorithm
DNN	Deep Neural Networks
DP	Differential Privacy
DRL	Deep Reinforcement Learning
FedAvg	Federated Averaging
FedZO	Federated Zeroth-Order Optimization
FEEL	Federated Edge Learning
GD	Gradient Descent
GNN	Graph Neural Network
GS	Gibbs Sampling
IoT	Internet of Things
KKT	Karush-Kuhn-Tucker
L-BFGS	Limited-Memory Broyden-Fletcher-Goldfarb-Shanno
LoS	Line-of-Sight
MEC	Mobile Edge Computing
ML	Machine Learning
MLP	Multi-Layer Perceptron
MSE	Mean-Squared-Error
NLoS	Non-Line-of-Sight
PBFT	Practical Byzantine Fault Tolerance

PoS	Proof of Stake
PoW	Proof of Work
PS	Parameter Server
PSD	Positive Semidefinite
QCQP	Quadratically Constrained Quadratic Programming
ReLU	Rectified Linear Unit
RIS	Reconfigurable Intelligent Surface
SDR	Semidefinite Relaxation
SGD	Stochastic Gradient Descent
SNR	Signal-to-Noise Ratio
SOC	Second-Order Cone
SVM	Support Vector Machine
UAV	Unmanned Aerial Vehicle
ZO-SGD	Zeroth-Order Stochastic Gradient Descent

Part I
Introduction

Chapter 1
Introduction

1.1 Overview of FEEL

The exponential increase in data volume has driven the proliferation of artificial intelligence (AI) applications such as image recognition and natural language processing, which are made possible by recent advances in machine learning (ML) techniques, especially deep learning, and the advent of unprecedented computing power. Traditionally, ML processes, including both training and inference, have been supported by cloud computing–centralized cloud data centers offering extensive access to computation, storage, and datasets. However, the rise of intelligent mobile devices and critical applications such as drones, smart vehicles, and augmented reality, necessitate low latency and high privacy, rendering cloud-based ML methodologies inadequate. Consequently, it has become increasingly appealing to keep data local to edge devices and perform training/inference directly at the edge, bypassing the need to transmit data to the cloud or networks. This emerging approach is known as edge AI [1]. The primary challenge lies in the limited computational, storage, energy, and bandwidth resources available to support mobile edge intelligent services. To tackle this issue, recent research has focused on reducing storage overhead, time, and power consumption during inference through model compression methods via hardware and software co-design [2]. Additionally, advanced distributed optimization algorithms [3–5] have been developed to accelerate the training process by leveraging the computing power and distributed data across multiple devices.

Federated edge learning (FEEL) [5–8] explores the possibility of distributed learning directly on mobile devices to enhance privacy and reduce network bandwidth usage. FEEL leverages the computational capabilities of edge devices to collaboratively train models while keeping the data localized. This approach addresses several challenges associated with traditional centralized ML approaches, including data privacy, latency, and bandwidth constraints. By distributing the

Y. Zhou et al., *Federated Edge Learning*, Wireless Networks,
https://doi.org/10.1007/978-3-031-96649-1_1

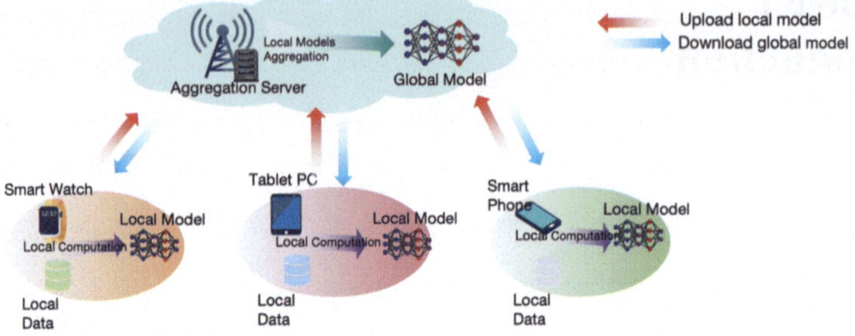

Fig. 1.1 FEEL over wireless networks

learning process across multiple edge devices, FEEL aims to create efficient and privacy-preserving machine learning models.

FEEL involves training ML models across distributed edge devices or servers holding local data samples without exchanging them. The architecture of FEEL typically includes edge devices, a central server, and a communication network. Edge devices such as smartphones, Internet of things (IoT) sensors, and smart home devices, each training a local model on its own data. The FEEL process begins with the central server initializing the global model and sending it to all participating edge devices. Each edge device then trains the model on its local data for a fixed number of iterations or epochs, generating local model updates (e.g., weight updates or gradients). These local updates are sent to the central server, which aggregates them using techniques like Federated Averaging (FedAvg) to update the global model. The updated global model is then distributed back to the edge devices, and this process repeats iteratively until the model converges or a stopping criterion is met. The communication network facilitates communication between edge devices and the central server, with efficient communication protocols being essential to minimize bandwidth usage and latency. The training process of FEEL is illustrated in Fig. 1.1. In summary, FEEL represents a significant advancement in the field of distributed ML, leveraging local computational resources and ensuring data privacy, FEEL offers a scalable, efficient, and privacy-preserving approach to training AI models over wireless networks.

1.2 Learning Models and Algorithms in FEEL

The models utilized in FEEL are consistent with those used in the conventional ML. Before diving into the core concept of this book, we provide an overview of the fundamental ML models and algorithms that form the basis of our following chapters. In this section, we introduce these basic models and algorithms, which shall help readers understand the subsequent chapters of this book.

1.2.1 Basics of ML Model

Logistic Regression Logistic regression is a classical model that is primarily employed to address binary classification tasks. Its fundamental objective involves estimating the probability of an input example belonging to the "1" class as opposed to the "0" class. In a more concrete sense, logistic regression endeavors to acquire knowledge of a function with the following structure:

$$
\begin{aligned}
P(y=1 \mid \boldsymbol{x}) &= h_\theta(\boldsymbol{x}) = \frac{1}{1+\exp\left(-\boldsymbol{\theta}^\top \boldsymbol{x}\right)} \equiv \sigma\left(\boldsymbol{\theta}^\top \boldsymbol{x}\right), \\
P(y=0 \mid \boldsymbol{x}) &= 1 - P(y=1 \mid \boldsymbol{x}) = 1 - h_\theta(\boldsymbol{x}).
\end{aligned} \tag{1.1}
$$

In this context, $\boldsymbol{x}$ denotes the feature vector of a sample, while y signifies its label. The function $\sigma(z) \equiv \frac{1}{1+\exp(-z)}$ is commonly referred to as the "sigmoid" or "logistic" function. It exhibits an S-shaped curve, effectively "compressing" the value of $\theta^\top \boldsymbol{x}$ into the interval (0, 1), allowing us to interpret $h_\theta(\boldsymbol{x})$ as a probability.

The main objective is to determine an appropriate value for θ such that the probability $P(y = 1 \mid \boldsymbol{x}) = h_\theta(\boldsymbol{x})$ is significant when $\boldsymbol{x}$ pertains to the "1" class and minimal when $\boldsymbol{x}$ pertains to the "0" class (ensuring that $P(y = 0 \mid \boldsymbol{x})$ is high). When dealing with a set of training examples equipped with binary labels, denoted as $\left\{\left(\boldsymbol{x}^{(i)}, y^{(i)}\right) : i = 1, \ldots, m\right\}$, we gauge the performance of a given h_θ using the following cost function:

$$
J(\boldsymbol{\theta}) = -\sum_i \left(y^{(i)} \log\left(h_\theta\left(\boldsymbol{x}^{(i)}\right)\right) + \left(1 - y^{(i)}\right) \log\left(1 - h_\theta\left(\boldsymbol{x}^{(i)}\right)\right)\right). \tag{1.2}
$$

Note that only one of the two terms within the summation contributes to each training example, depending on the value of the label $y^{(i)}$ (whether it's 0 or 1). Specifically, when $y^{(i)} = 1$, minimizing the cost function entails increasing $h_\theta\left(\boldsymbol{x}^{(i)}\right)$, whereas when $y^{(i)} = 0$, we strive to maximize $1 - h_\theta$ as previously elucidated.

The aforementioned cost function serves as a gauge for how effectively a given hypothesis h_θ aligns with our training data. Our aim is to learn the classification of our training data by minimizing $J(\boldsymbol{\theta})$, effectively discovering the optimal choice of θ. Once this is achieved, we can classify new test points as either "1" or "0" by assessing which of these class labels is more probable: if $P(y = 1 \mid \boldsymbol{x}) > P(y = 0 \mid \boldsymbol{x})$, we label the example as "1"; otherwise, it's labeled as "0". This aligns with the condition of checking whether $h_\theta(x) > 0.5$.

In order to minimize $J(\boldsymbol{\theta})$, we can employ the gradient descent method [9]. We must supply a function that calculates both $J(\boldsymbol{\theta})$ and $\nabla_\theta J(\boldsymbol{\theta})$ for any specified choice of $\boldsymbol{\theta}$. The derivative of $J(\boldsymbol{\theta})$, as presented earlier, with respect to the j-th element of $\boldsymbol{\theta}$ denoted as θ_j is as follows:

$$\frac{\partial J(\boldsymbol{\theta})}{\partial \theta_j} = \sum_i \boldsymbol{x}_j^{(i)} \left(h_\theta \left(\boldsymbol{x}^{(i)} \right) - y^{(i)} \right). \tag{1.3}$$

In its vectorized representation, the complete gradient can be formulated as follows:

$$\nabla_\theta J(\boldsymbol{\theta}) = \sum_i \boldsymbol{x}^{(i)} \left(h_\theta \left(\boldsymbol{x}^{(i)} \right) - y^{(i)} \right). \tag{1.4}$$

Softmax Regression Softmax regression, also known as multinomial logistic regression, represents an extension of logistic regression tailored to the scenario where we need to handle multiple classes [10]. In traditional logistic regression, we utilize binary labels: $y^{(i)} \in \{0, 1\}$, typically utilized for tasks like distinguishing between two types of handwritten digits. However, softmax regression empowers us to address a broader spectrum of labels: $y^{(i)} \in \{1, \ldots, K\}$, where K denotes the number of classes.

Recalling logistic regression, we operated within the framework of binary classification, and our training set comprised m labeled examples with input features $\boldsymbol{x}^{(i)} \in \mathbb{R}^n$. The hypothesis takes the following form

$$h_\theta(\boldsymbol{x}) = \frac{1}{1 + \exp\left(-\boldsymbol{\theta}^\top \boldsymbol{x}\right)}, \tag{1.5}$$

with model parameters $\boldsymbol{\theta}$ trained to minimize the cost function:

$$J(\boldsymbol{\theta}) = -\left[\sum_{i=1}^{m} y^{(i)} \log h_\theta \left(\boldsymbol{x}^{(i)} \right) + \left(1 - y^{(i)} \right) \log \left(1 - h_\theta \left(\boldsymbol{x}^{(i)} \right) \right) \right]. \tag{1.6}$$

In the context of softmax regression, our objective shifts toward multi-class classification, accommodating more than just two classes. Consequently, in the training set $\{(\boldsymbol{x}^{(1)}, y^{(1)}), \ldots, (\boldsymbol{x}^{(m)}, y^{(m)})\}$, we now observe that $y^{(i)} \in \{1, 2, \ldots, K\}$. For instance, in tasks like MNIST digit recognition, $K = 10$. It's important to note that we adopt the convention of indexing classes starting from 1.

For a given test input $\boldsymbol{x}$, our objective is to have our hypothesis provide estimations for the probabilities $P(y = k \mid \boldsymbol{x})$ across all values of $k = 1, \ldots, K$. Essentially, we aim to assess the likelihood of the class label adopting any of the K distinct values. As a result, our hypothesis will generate a K-dimensional vector, with the properties that its elements sum up to 1, thereby representing our estimated probabilities for each of the K potential outcomes. More explicitly, our hypothesis, denoted as $h_\theta(\boldsymbol{x})$, is formulated as follows:

$$h_\theta(\boldsymbol{x}) = \begin{bmatrix} P(y=1 \mid \boldsymbol{x};\boldsymbol{\theta}) \\ P(y=2 \mid \boldsymbol{x};\boldsymbol{\theta}) \\ \vdots \\ P(y=K \mid \boldsymbol{x};\boldsymbol{\theta}) \end{bmatrix} = \frac{1}{\sum_{k=1}^{K} \exp\left(\boldsymbol{\theta}^{(k)\top}\boldsymbol{x}\right)} \begin{bmatrix} \exp\left(\boldsymbol{\theta}^{(1)\top}\boldsymbol{x}\right) \\ \exp\left(\boldsymbol{\theta}^{(2)\top}\boldsymbol{x}\right) \\ \vdots \\ \exp\left(\boldsymbol{\theta}^{(K)\top}\boldsymbol{x}\right) \end{bmatrix}.$$

Here, the model parameters are denoted as $\boldsymbol{\theta}^{(1)}, \boldsymbol{\theta}^{(2)}, \ldots, \boldsymbol{\theta}^{(K)} \in \mathbb{R}^n$. Note that term $\frac{1}{\sum_{k=1}^{K} \exp\left(\boldsymbol{\theta}^{(k)\top}\boldsymbol{x}\right)}$ serves to normalize the distribution, ensuring that it sums up to one.

Next, we outline the cost function used in softmax regression. In the equation below, we utilize the indicator function $\mathbf{1}\{\cdot\}$, which evaluates to 1 for a true statement and 0 for a false statement. For instance, $\mathbf{1}\{1+1=2\}$ equals 1, while $\mathbf{1}\{1+1=3\}$ equals 0. The cost function is defined as follows [11]:

$$J(\boldsymbol{\theta}) = -\left[\sum_{i=1}^{m}\sum_{k=1}^{K} \mathbf{1}\left\{y^{(i)} = k\right\} \log \frac{\exp\left(\boldsymbol{\theta}^{(k)\top}\boldsymbol{x}^{(i)}\right)}{\sum_{k=1}^{K} \exp\left(\boldsymbol{\theta}^{(k)\top}\mathbf{x}^{(i)}\right)}\right]. \tag{1.7}$$

Notice that this extends the logistic regression cost function, which could also be expressed as:

$$\begin{aligned} J(\boldsymbol{\theta}) &= -\left[\sum_{i=1}^{m}\left(1-y^{(i)}\right)\log\left(1-h_\theta\left(\boldsymbol{x}^{(i)}\right)\right) + y^{(i)}\log h_\theta\left(\boldsymbol{x}^{(i)}\right)\right] \\ &= -\left[\sum_{i=1}^{m}\sum_{k=0}^{K} \mathbf{1}\left\{y^{(i)}=k\right\}\log P\left(y^{(i)}=k \mid \boldsymbol{x}^{(i)};\boldsymbol{\theta}\right)\right]. \end{aligned} \tag{1.8}$$

The softmax cost function follows a similar structure, with the difference being that we now sum over the k possible class label values. Note that in softmax regression, we have the probability expression:

$$P\left(y^{(i)}=k \mid \boldsymbol{x}^{(i)};\boldsymbol{\theta}\right) = \frac{\exp\left(\boldsymbol{\theta}^{(k)\top}\boldsymbol{x}^{(i)}\right)}{\sum_{k=1}^{K}\exp\left(\boldsymbol{\theta}^{(k)\top}\boldsymbol{x}^{(i)}\right)}. \tag{1.9}$$

Solving for the minimum of $J(\boldsymbol{\theta})$ analytically is not feasible, so we rely on iterative optimization algorithms. By taking derivatives, we can compute the gradient as follows:

$$\nabla_{\theta^{(k)}} J(\boldsymbol{\theta}) = -\sum_{i=1}^{m}\left[\boldsymbol{x}^{(i)}\left(1\left\{y^{(i)}=k\right\} - P\left(y^{(i)}=k \mid \boldsymbol{x}^{(i)};\theta\right)\right)\right]. \tag{1.10}$$

Here, $\nabla_{\theta^{(k)}} \boldsymbol{J}(\theta)$ is a vector, and its j-th element represents the partial derivative of $J(\boldsymbol{\theta})$ concerning the j-th element of $\boldsymbol{\theta}^{(k)}$. Utilizing this derivative, we can find the optimal $\boldsymbol{\theta}$ that minimizes $J(\boldsymbol{\theta})$ through standard optimization techniques, such as gradient descent.

Multi-Layer Perceptron Multi-layer perceptron (MLP) serves as a straightforward neural network that acquires the ability to learn a function $f : \mathbb{R}^n \rightarrow \mathbb{R}^d$ through the process of training on a dataset $\{(\boldsymbol{x}^{(1)}, y^{(1)}), \ldots, (\boldsymbol{x}^{(m)}, y^{(m)})\}$, encompassing m labeled samples [12]. In this context, n signifies the input dimensionality, while d represents the output dimensionality. When provided with a set of features $\{\boldsymbol{x}^{(i)}\}$ and a corresponding target set $\{y^{(i)}\}$, the MLP has the capability to learn a non-linear function approximator, a feature useful for both classification and regression tasks. Notably, the MLP distinguishes itself from softmax regression by the inclusion of one or more non-linear intermediary layers, often referred to as hidden layers, situated between the input and output layers.

In Fig. 1.2, we can observe a two-hidden layer MLP designed for a three-category classification task. The initial layer on the left, referred to as the input layer, comprises a collection of neurons that represent the input features. Specifically, each component of $\boldsymbol{x}_j$ corresponds to an individual feature. Within the hidden layer, each neuron undertakes a transformation of values from the preceding layer using a weighted linear summation operation, followed by the application of a non-linear activation function denoted as $\sigma : \mathbb{R} \rightarrow \mathbb{R}$. It's noteworthy that this activation

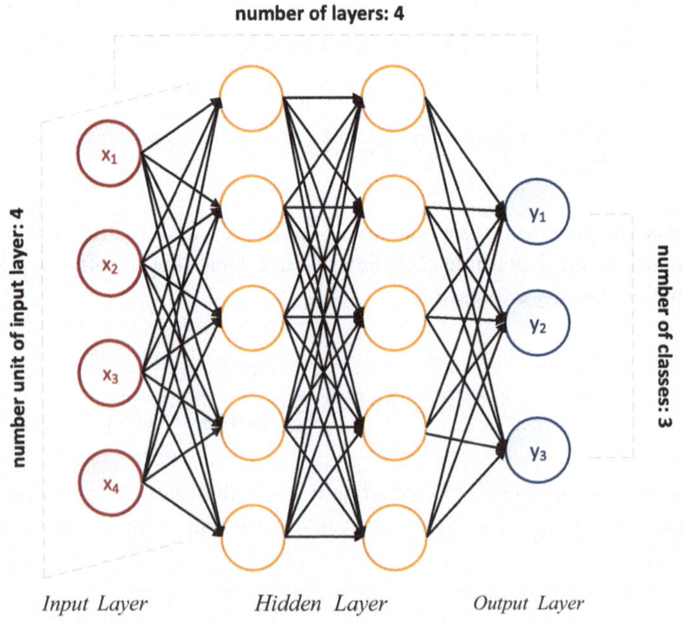

Fig. 1.2 Architecture of a multi-layer perceptron with 2 hidden layers

Table 1.1 Some basic activation functions

Name	Formulation
Sigmoid	$\sigma(x) = \frac{1}{1+e^{-x}}$
Tanh	$\sigma(x) = \frac{e^x - e^{-x}}{e^x + e^{-x}}$
ReLU	$\sigma(x) = \max(0, x)$
Exponential linear unit	$\sigma(x) = \begin{cases} x & \text{if } x > 0, \\ \alpha(e^x - 1) & \text{otherwise} \end{cases}$
Softmax	$\sigma_i(x) = \frac{e^{x_i}}{\sum_{j=1}^{d} e^{x_j}}$ for $i = 1, \ldots, d$
Maxout	$\sigma(x) = \max_i x_i$

function could take various forms, including ReLU, sigmoid, or tanh functions, as detailed in Table 1.1 [13]. Subsequently, the output layer receives these values from the last hidden layer and performs further transformations to produce the final output values.

In mathematical terms, we can represent an MLP with r hidden neurons and a one-dimensional output as follows [14]:

$$\hat{y} = \phi\left(\boldsymbol{m}^\top \sigma\left(\boldsymbol{W}^\top \boldsymbol{x} + \boldsymbol{b}_1\right) + b_2\right).$$

Here, $\boldsymbol{x} \in \mathbb{R}^n$ denotes the input vector, representing the sample features. The weights and bias between the input layer neurons and hidden layer neurons are denoted by $\boldsymbol{W} \in \mathbb{R}^{n \times r}$ and $\boldsymbol{b} \in \mathbb{R}^r$, respectively. All hidden layer neurons share the same activation function $\sigma(\cdot)$. The output value $\hat{y}$ is generated using the weights $\boldsymbol{m} \in \mathbb{R}^r$ and bias b_2, and it is transformed by the activation function $\phi(\cdot)$.

The MLP model can effectively perform various classification and regression tasks when the weights and biases are learned accurately. The process of training the neural network is transformed into an optimization problem, aiming to minimize the loss function by adjusting the weights and bias. Let's consider the square loss as an example. The corresponding loss function for the dataset $\{(\boldsymbol{x}^{(1)}, y^{(1)}), \ldots, (\boldsymbol{x}^{(m)}, y^{(m)})\}$ is defined as:

$$\mathcal{L}(\boldsymbol{W}, \boldsymbol{b}_1, \boldsymbol{m}, b_2) = \sum_i \left(y^{(i)} - \phi\left(\boldsymbol{m}^\top \sigma\left(\boldsymbol{W}^\top \boldsymbol{x}^{(i)} + \boldsymbol{b}_1\right) + b_2\right)\right)^2.$$

Consequently, the primary challenge lies in finding the optimal weights and bias, which involves solving the following optimization problem:

$$\underset{\boldsymbol{W}, \boldsymbol{b}_1, \boldsymbol{m}, b_2}{\text{minimize}} \mathcal{L}(\boldsymbol{W}, \boldsymbol{b}_1, \boldsymbol{m}, b_2).$$

This optimization problem can be efficiently solved using standard optimization algorithms, where the gradients can be computed using the well-established backward-propagation mechanism [15].

Convolutional Neural Networks Convolutional neural networks (CNNs) [16] share many similarities with regular neural networks discussed in the preceding section. They comprise neurons with adjustable weights and biases, each receiving inputs, conducting a dot product operation, and potentially applying a non-linear transformation. The entire network continues to represent a unified differentiable score function, spanning from the initial raw image pixel inputs to the final class scores. Furthermore, they retain a loss function, such as SVM or Softmax, in the last (fully connected) layer, and all the methodologies we devised for training standard Neural Networks remain applicable.

CNN architectures operate on the implicit assumption that their inputs constitute images, enabling the integration of specific characteristics into the structure. Consequently, this simplifies the implementation of the forward function and leads to a significant reduction in the network's parameter count.

The fundamental building block of a convolutional network, known as the convolutional layer, performs the majority of the computationally intensive tasks. The parameters within the convolutional layer consist of a collection of learnable filters. These filters are relatively compact in terms of spatial dimensions, typically featuring dimensions like $5 \times 5 \times 3$ (representing 5 pixels in width and height, with a depth of 3 to accommodate color channels). Consider the forward pass, during which we perform a sliding operation, more precisely, a convolution, of each filter across the input volume's width and height. This convolution computes dot products between the filter's elements and the input at each position, resulting in the generation of a 2-dimensional activation map that captures the filter's responses at various spatial positions.

Within each convolutional layer, it's customary to find a set of filters. Each filter independently produces its own distinct 2-dimensional activation map. These activation maps are then stacked along the depth dimension to construct the final output volume for that layer [17].

As shown in Fig. 1.3, the CNN architecture is composed of a sequence of convolutional and subsampling layers, which may optionally be followed by fully connected layers. The input to a convolutional layer takes the form of a $d \times d \times r$ image, where d represents the image's height and width, and r signifies the number of channels (e.g., the above figure corresponding to $r = 3$ channels). Within a

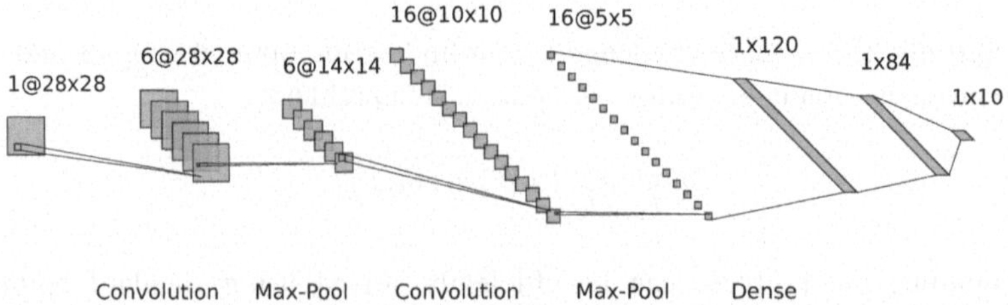

Fig. 1.3 Architecture of a convolutional neural network

convolutional layer, there exist k filters, or kernels, each with dimensions of $n \times n \times q$. Notably, n is typically smaller than the image's dimension, and q can be either equal to the number of channels r or smaller, with the possibility of varying for each kernel. The filter sizes contribute to the locally connected structure, as each filter convolves with the input image, resulting in the creation of k feature maps, each sized $d - n + 1$. Subsequently, these maps typically undergo subsampling, often through mean or max pooling, conducted over contiguous regions of size $p \times p$, where p is typically set to 2 for smaller images (e.g., MNIST) and generally remains below 5 for larger input sizes. An additive bias and sigmoidal nonlinearity may be applied to each feature map, either before or after the subsampling layer.

Following the convolutional layers, the architecture may include any number of dense connected layers, which closely resemble the layers in a conventional multi-layer perceptron. The computation of the loss function (e.g., Softmax) relies on the output produced by the final fully connected layer.

1.2.2 Preliminary of Optimization Algorithms

From the perspective of optimization, the objective of the model training is to solve an unconstrained optimization problem

$$\underset{\boldsymbol{x}}{\text{minimize}} f(\boldsymbol{x}). \tag{1.11}$$

To solve this problem, many algorithms have been proposed, ranging from first-order, second-order, and zeroth-order algorithms. In FEEL, such a problem will be presented in a distributed fashion. In this subsection, we will mainly focus on the centralized setting and introduce some conventional algorithms for solving problem (1.11), which would lay the foundation of Part II.

Gradient Based Optimization Gradient descent (GD) is the most representative algorithm for solving problem (1.11). It relies only on gradient information, which categorizes it as a first-order method. The core idea of gradient descent is to iteratively update the parameters in the direction opposite to the gradient of the objective function. The update rule for gradient descent is given by:

$$\boldsymbol{x}_{t+1} = \boldsymbol{x}_t - \eta \nabla f(\boldsymbol{x}_t), \tag{1.12}$$

where $\boldsymbol{x}_t$ is the parameter vector at iteration t, η is the learning rate, and $\nabla f(\boldsymbol{x}_t)$ is the gradient of the objective function at $\boldsymbol{x}_t$. Gradient descent is simple to implement and computationally efficient, making it a popular choice for many machine-learning problems [9, 18].

Several variants of gradient descent have been developed to improve its performance and convergence properties, such as stochastic gradient descent (SGD). As an extension of GD in random scenarios, SGD was proved to be more effective for large-scale machine learning model training [18]. Instead of using the full gradient, SGD uses a randomly selected subset (mini-batch) of the data to compute the gradient, resulting in faster updates and the ability to handle large datasets. Actually, SGD has become a popular optimization method for training deep learning models due to its efficiency and scalability. The update rule for SGD is given by:

$$\boldsymbol{x}_{t+1} = \boldsymbol{x}_t - \eta \nabla f_i(\boldsymbol{x}_t), \tag{1.13}$$

where $\nabla f_i(\boldsymbol{x}_t)$ is the gradient computed from the i-th randomly selected data point or mini-batch. This introduces noise into the optimization process, which helps in escaping local minima and exploring the parameter space more effectively. Despite its stochastic nature, SGD can converge to a good approximation of the global minimum, especially with proper tuning of the learning rate and other hyperparameters.

Hessian Based Optimization—Newton Algorithm While the first-order algorithms, like GD, can solve problem (1.11), their convergence speed is generally slow due to the limitation of gradient information. It can be attributed to the fact that gradient information doesn't capture the curvature of this problem. To tackle this challenge and accelerate the convergence speed. Newton's method was proposed. As a second-order optimization algorithm, Newton's method uses both the gradient and the Hessian matrix (second-order derivatives) of the objective function. The update rule for Newton's method is:

$$\boldsymbol{x}_{t+1} = \boldsymbol{x}_t - \eta \boldsymbol{H}^{-1} \nabla f(\boldsymbol{x}_t), \tag{1.14}$$

where $\boldsymbol{H}$ is the Hessian matrix of the objective function at $\boldsymbol{x}_t$.

To be specific, the updating rule can be explained as minimizing the second-order Taylor expansion of the objective function at point $\boldsymbol{x}_t$ [19]. Concretely, the second-order Taylor expansion of the objective function $f(\boldsymbol{x})$ at point $\boldsymbol{x}_t$ is

$$f(\boldsymbol{x}) \approx f(\boldsymbol{x}_t) + \nabla f(\boldsymbol{x}_t)(\boldsymbol{x} - \boldsymbol{x}_t) + \frac{1}{2}(\boldsymbol{x} - \boldsymbol{x}_t)^\top \nabla^2 f(\boldsymbol{x}_t)(\boldsymbol{x} - \boldsymbol{x}_t),$$

where $\nabla^2 f(\boldsymbol{x}_t)$ is called the Hessian matrix of $f(\boldsymbol{x}_t)$, i.e., $\boldsymbol{H}$. For convex function, H is always a positive semi-definite matrix. The minimizer for the approximation function is thus $\boldsymbol{x}_t - \eta \boldsymbol{H}^{-1} \nabla f(\boldsymbol{x}_t)$. In other words, the update rule of Newton method is just minimizing the second-order Taylor expansion of the objective function $f(\boldsymbol{x})$ at the current iterate.

Based on the idea of Newton's method, many other second-order methods using Hessian information have been proposed, such as Broyden-Fletcher-Goldfarb-Shanno (BFGS) [20] algorithm and its limited-memory variant (L-BFGS). In summary, Newton-type methods can achieve faster convergence than first-order methods, especially near the optimum, due to their use of second-order information. A well-recognized reason is that the combination of gradient and Hessian information reflects the local geometry of this objective function better than the sole use of gradient. However, the implementation of a second-order algorithm is actually more challenging than the first-order algorithms. Concretely, computing and inverting the Hessian matrix can be computationally expensive. There is a trade-off between convergence rate and communication efficiency. Notably, the criteria for centralized and federated settings are slightly different. Due to limited radio frequency resources, the pursuit of convergence rate is more pronounced in FEEL, making second-order algorithms more appealing.

Gradient-Free Optimization—Zeroth-Order Algorithm First-order algorithms shine with their easy implementation while second-order methods excel in achieving a fast convergence rate. However, these two different types of algorithms rely on either gradient or hessian information. However, in some scenarios, gradient information may not be available or is too costly to compute. This can occur in cases where the objective function is noisy, discontinuous, or only available as a black-box function. Zeroth-order optimization, also known as gradient-free optimization, offers a solution solely using function value queries to guide the search for the optimum. These methods are essential in various fields, including hyperparameter tuning, black-box optimization, and simulation-based optimization, where the gradient and Hessian information are not accessible.

A commonly used strategy for the zeroth-order algorithm is to use finite difference methods to approximate the gradient of an objective function. By evaluating the objective function at several nearby points, finite difference methods provide an estimate of the gradient, which can then be used to guide optimization algorithms. In particular, the central difference formula is one of the most commonly used finite difference methods for gradient approximation. It provides a more accurate estimate of the gradient compared to other finite difference approaches. The central difference formula for the partial derivative of a function f with respect to the i-th variable x_i is given by Liu et al. [21]:

$$\frac{\partial f}{\partial x_i} \approx \frac{f(\boldsymbol{x} + h\boldsymbol{e}_i) - f(\boldsymbol{x} - h\boldsymbol{e}_i)}{2h}$$

where $\boldsymbol{x}$ is the point at which the gradient is being approximated, h is a small perturbation or step size, and $\boldsymbol{e}_i$ is the unit vector in the i-th dimension, which means that only the i-th component of $\boldsymbol{e}_i$ is 1 and all other components are 0.

This formula works by evaluating the function f at two points: one slightly ahead of and one slightly behind the point of interest in the i-th dimension. The difference between these function values, divided by twice the step size, approximates the gradient. We then replace the true gradient in first-order algorithms with the approximated gradient, resulting in zeroth-order algorithms [22]. For FEEL, what remains to do is to extend the centralized zeroth-order algorithm to the distributed setup, which will be discussed in Part II.

1.3 Challenges of FEEL

The rapid advancement of FEEL is driven by the need for enhanced data privacy, low communication and computation latency, as well as efficient utilization of communication and computation resources at the network edge. However, the implementation of FEEL also presents several challenges that need to be addressed for its successful deployment in practice.

Algorithmic Convergence over Wireless Networks With conventional orthogonal multiple access schemes, the sequential communication of high-dimensional model updates between numerous edge devices and the central server introduces excessive delays. To solve this issue, we shall introduce over-the-air computation (AirComp) assisted model aggregation to support communication-efficient FEEL [23]. By employing AirComp, we aim to aggregate local updates concurrently, thereby minimizing the communication rounds and expediting the overall learning process. This concurrent aggregation is particularly beneficial in scenarios with a large number of participating devices, where communication overhead is a critical bottleneck. The primary challenge is to ensure algorithmic convergence in the presence of random channel fading and receiver noise. In addition, wireless transmission inherently suffers from co-channel interference, which can distort the transmitted updates and affect the accuracy of the aggregated model. Besides, fading channels can introduce variability in signal strength, further complicating the communication process. To address these issues, rigorous convergence analysis is required, along with the procedures to mitigate the impact of noise and channel fading on the learning process [24, 25].

Communication Bottleneck Although AirComp has the potential to significantly reduce the communication latency, the achievable model aggregation accuracy is constrained by the device with the worst channel condition toward the edge server, leading to the severe communication bottleneck issue that limits the learning performance. This is because, according to the principle of AirComp, all devices should adjust their transmit power to ensure that the signal powers received at the edge server are aligned. However, due to random channel fading, the channel qualities between the edge server and edge devices are diverse. With the same maximum transmit power constraint, the device with the worst channel condition toward the edge server limits the power magnitude that can be aligned, which becomes the communication bottleneck. To tackle the communication bottleneck issue, reconfigurable intelligent surfaces (RIS) can be utilized to manipulate the propagation environment by dynamically adjusting the phase-shifters to enhance the channel quality of the worst channel, thereby mitigating the communication bottleneck issue [26]. In addition, by exploiting the high altitude and mobility, unmanned aerial vehicles (UAVs) offer unique advantages of establishing short-distance line-of-sight links with all edge devices, which prevents any single device from becoming a communication bottleneck, thereby enhancing the overall reliability and performance of the FEEL system [27]. Optimizing the placement and configuration of RIS and UAVs requires sophisticated algorithms that can balance multiple factors, such as signal quality, coverage, and energy consumption. Managing the energy resources of these devices is also critical, as their operational lifespan directly impacts the sustainability and efficiency of the FEEL system. In addition, the interference management among multiple cells also raises unique challenge.

Security and Privacy Even though the local raw data of edge devices are not shared in FEEL, the exchange of local model updates may still leak sensitive information. To further alleviate the privacy concern, differential privacy (DP) emerges as a promising mechanism that introduces controllable artificial noise to the local model updates, ensuring that individual data points cannot be inferred from the aggregated model. This mechanism is capable of providing strong privacy guarantees and suitable for applications with high sensitivity requirement, such as healthcare and finance. On the other hand, mitigating the impact of malicious users is another important aspect for FEEL. The main challenge in implementing these security and privacy mechanisms is balancing the trade-off among privacy, security, and learning performance. Specifically, the DP mechanisms, while providing strong privacy guarantees, introduces additional artificial noise that may reduce the model aggregation accuracy and degrade the learning performance [28]. Additionally, Byzantine-resilient aggregation techniques must be robust enough to detect and mitigate malicious updates without excessively filtering out legitimate updates, which could hinder the learning process [29]. Achieving such a balance requires careful design and extensive evaluation of these techniques to ensure they can provide the desired level of security and privacy without compromising on the learning performance.

1.4 Organization

This book is organized as follows: In the second part, we discussed different algorithms for FEEL, ranging from first-order, second-order, and zeroth-order algorithms. In the third part, we present FEEL under different system architectures. Specifically, we introduce RIS-assisted FEEL, UAV-assisted FEEL, and multi-cell FEEL. In the fourth part, we shift our focus to the privacy and security issues in FEEL and present DP model aggregation and Byzantine-resilient model aggregation mechanisms.

References

1. Y. Shi, K. Yang, T. Jiang, J. Zhang, and K. B. Letaief, "Communication-efficient edge AI: Algorithms and systems," *IEEE Communications Surveys & Tutorials*, vol. 22, no. 4, pp. 2167–2191, 2020.
2. S. Han, H. Mao, and W. J. Dally, "Deep compression: Compressing deep neural networks with pruning, trained quantization and huffman coding," *arXiv preprint arXiv:1510.00149*, 2015.
3. Y. Lin, S. Han, H. Mao, Y. Wang, and W. J. Dally, "Deep gradient compression: Reducing the communication bandwidth for distributed training," *arXiv preprint arXiv:1712.01887*, 2017.
4. C. Karakus, Y. Sun, S. Diggavi, and W. Yin, "Straggler mitigation in distributed optimization through data encoding," *Advances in Neural Information Processing Systems*, vol. 30, 2017.
5. B. McMahan, E. Moore, D. Ramage, S. Hampson, and B. A. y Arcas, "Communication-efficient learning of deep networks from decentralized data," in *Artificial Intelligence and Statistics*, pp. 1273–1282, PMLR, 2017.
6. Y. Zhou, Y. Shi, H. Zhou, J. Wang, L. Fu, and Y. Yang, "Toward scalable wireless federated learning: Challenges and solutions," *IEEE Internet Things Mag.*, vol. 6, pp. 10–16, Apr. 2023.
7. K. Yang, Y. Shi, Y. Zhou, Z. Yang, L. Fu, and W. Chen, "Federated machine learning for intelligent IoT via reconfigurable intelligent surface," *IEEE network*, vol. 34, pp. 16–22, May 2020.
8. Z. Wang, Y. Zhou, Y. Shi, and W. Zhuang, "Interference management for over-the-air federated learning in multi-cell wireless networks," *IEEE J. Sel. Areas Commun.*, vol. 40, pp. 2361–2377, Aug. 2022.
9. S. P. Boyd and L. Vandenberghe, *Convex optimization*. Cambridge university press, 2004.
10. E. L. Allwein, R. E. Schapire, and Y. Singer, "Reducing multiclass to binary: A unifying approach for margin classifiers," *Journal of machine learning research*, vol. 1, no. Dec, pp. 113–141, 2000.
11. C. M. Bishop and N. M. Nasrabadi, *Pattern recognition and machine learning*, vol. 4. Springer, 2006.
12. J. Schmidhuber, "Deep learning in neural networks: An overview," *Neural networks*, vol. 61, pp. 85–117, 2015.
13. G. Hinton, L. Deng, D. Yu, G. E. Dahl, A.-r. Mohamed, N. Jaitly, A. Senior, V. Vanhoucke, P. Nguyen, T. N. Sainath, *et al.*, "Deep neural networks for acoustic modeling in speech recognition: The shared views of four research groups," *IEEE Signal processing magazine*, vol. 29, no. 6, pp. 82–97, 2012.
14. I. Goodfellow, Y. Bengio, and A. Courville, *Deep learning*. MIT press, 2016.
15. J. Schmidhuber, "Annotated history of modern ai and deep learning," *arXiv preprint arXiv:2212.11279*, 2022.
16. Y. LeCun, B. Boser, J. S. Denker, D. Henderson, R. E. Howard, W. Hubbard, and L. D. Jackel, "Backpropagation applied to handwritten zip code recognition," *Neural computation*, vol. 1, no. 4, pp. 541–551, 1989.
17. M. D. Zeiler and R. Fergus, "Visualizing and understanding convolutional networks," in *Computer Vision–ECCV 2014: 13th European Conference, Zurich, Switzerland, September 6–12, 2014, Proceedings, Part I 13*, pp. 818–833, Springer, 2014.
18. L. Bottou, F. E. Curtis, and J. Nocedal, "Optimization methods for large-scale machine learning," *SIAM review*, vol. 60, no. 2, pp. 223–311, 2018.
19. J. Nocedal and S. J. Wright, *Numerical optimization*. Springer, 1999.
20. C.-X. Kou and Y.-H. Dai, "A modified self-scaling memoryless broyden–fletcher–goldfarb–shanno method for unconstrained optimization," *Journal of Optimization Theory and Applications*, vol. 165, pp. 209–224, 2015.
21. S. Liu, P.-Y. Chen, B. Kailkhura, G. Zhang, A. O. Hero III, and P. K. Varshney, "A primer on zeroth-order optimization in signal processing and machine learning: Principals, recent advances, and applications," *IEEE Signal Processing Magazine*, vol. 37, no. 5, pp. 43–54, 2020.

22. Y. Nesterov and V. Spokoiny, "Random gradient-free minimization of convex functions," *Foundations of Computational Mathematics*, vol. 17, no. 2, pp. 527–566, 2017.
23. Z. Wang, Y. Zhao, Y. Zhou, Y. Shi, C. Jiang, and K. B. Letaief, "Over-the-air computation for 6G: Foundations, technologies, and applications," *IEEE Internet Things J.*, 2024.
24. W. Fang, Z. Yu, Y. Jiang, Y. Shi, C. N. Jones, and Y. Zhou, "Communication-efficient stochastic zeroth-order optimization for federated learning," *IEEE Trans. Sig. Process.*, vol. 70, pp. 5058–5073, 2022.
25. P. Yang, Y. Jiang, T. Wang, Y. Zhou, Y. Shi, and C. N. Jones, "Over-the-air federated learning via second-order optimization," *IEEE Trans. Wireless Commun.*, vol. 21, no. 12, pp. 10560–10575, 2022.
26. Z. Wang, J. Qiu, Y. Zhou, Y. Shi, L. Fu, W. Chen, and K. B. Letaief, "Federated learning via intelligent reflecting surface," *IEEE Transactions on Wireless Communications*, vol. 21, no. 2, pp. 808–822, 2022.
27. M. Fu, Y. Shi, and Y. Zhou, "Federated learning via unmanned aerial vehicle," *IEEE Trans. Wireless Commun.*, vol. 23, no. 4, pp. 2884–2900, 2024.
28. Y. Yang, Y. Zhou, Y. Wu, and Y. Shi, "Differentially private federated learning via reconfigurable intelligent surface," *IEEE Internet Things J.*, vol. 9, no. 20, pp. 19728–19743, 2022.
29. Z. Yang, Y. Shi, Y. Zhou, Z. Wang, and K. Yang, "Trustworthy federated learning via blockchain," *IEEE Internet Things J.*, vol. 10, no. 1, pp. 92–109, 2023.

Part II
Algorithms

In this part, we shall present various federated optimization algorithms, including zeroth-order, first-order, and second-order methods, over wireless networks, followed by leveraging over-the-air computation to support low-latency model aggregation. We shall also present the convergence analysis framework for each algorithm, discussing the impact of learning and communication parameters on the learning performance, as well as illustrating the convergence rate, computation complexity, and communication overhead of these algorithms.

Chapter 2
First-Order Algorithm for Federated Edge Learning

2.1 Overview

The conventional centralized ML workflow heavily relies on cloud computing infrastructure which is typically equipped with extensive computational power and storage capabilities, and has access to the entire dataset. However, the emergence of intelligent mobile devices and critical applications such as drones, smart vehicles, and augmented reality demands low-latency processing and enhanced privacy safeguards. Consequently, conventional cloud-based ML approaches, as outlined in [8], become impractical.

To address these evolving needs, a paradigm shift towards local data processing on edge devices is gaining traction. This paradigm, known as FEEL [11, 14], involves storing data directly on edge devices and conducting training and inference locally, thus eliminating the need to transmit data to centralized cloud or network resources. However, the primary challenge lies in the limited computational, storage, energy, and bandwidth resources available for mobile edge intelligent services. To tackle these limitations, a range of advanced distributed optimization algorithms [4, 9, 10, 13, 17] have been proposed. These algorithms leverage distributed data across multiple devices and exploit their computational capabilities to accelerate the training process. Among these approaches, the FedAvg algorithm [10] has emerged as a promising method for training ML models with reduced energy consumption.

As a first-order algorithm, FedAvg primarily relies on gradient information for local model updates, which is typically easy to obtain through automatic differentiation and backpropagation. This ease of gradient computation has contributed to the widespread popularity of first-order methods in the ML community. Additionally, compared to the conventional distributed stochastic gradient descent (DSGD) algorithm, FedAvg significantly reduces communication overhead, making it particularly attractive for the distributed training of large-scale machine learning models. Numerous studies have explored various aspects of FedAvg, including communication compression, model sparsification, and quantization. Additionally,

Y. Zhou et al., *Federated Edge Learning*, Wireless Networks,
https://doi.org/10.1007/978-3-031-96649-1_2

research has focused on improving its robustness to non-i.i.d. data distributions, enhancing its scalability, and optimizing its performance in resource-constrained environments. These advancements have further solidified FedAvg as a foundational algorithm in federated learning.

In the following chapter, we shall investigate the FedAvg algorithm and present a novel approach to speed up model aggregation within the FedAvg framework. Note that the global model aggregation process involves transmitting computed local updates from each device, followed by averaging computation at the central node. To achieve this, we develop a computation and communication co-design strategy, i.e., AirComp [1, 5, 19], and improve the performance and the convergence rate by jointly optimizing device selection and receive beamforming design. However, the joint design problem is a computationally challenging mixed combinatorial optimization problem with non-convex quadratic constraints. To address the limitations of existing algorithms for solving the presented sparse and low-rank optimization problem [16], we propose a unified DC approach to induce both the sparsity and low-rank structures. The proposed DC approach has the capability of accurately detecting the feasibility of non-convex quadratic constraints [12, 18]. We then develop a DC algorithm for the presented non-convex DC program via successive convex relaxation. The global convergence rate of the DC algorithm is further established by rewriting the DC function as the difference of strongly convex functions.

2.2 Federated First-Order Optimization

2.2.1 Federated Edge Learning

FEEL emerges as a new distributed learning paradigm which allows the geographically-dispersed edge devices collaboratively train a global model without data sharing [10]. Figure 2.1 illustrates a FEEL system with M single-antenna edge devices and one base station (BS) equipped with N antennas and powerful computing resources to support the following distributed ML task:

$$\underset{z\in\mathbb{R}^d}{\text{minimize}} \quad f(z) = \frac{1}{M}\sum_{m=1}^{M} f_m(z), \tag{2.1}$$

where z is the model parameter vector to be optimized with dimension d. Besides, $f_m(z)$ is the local loss function that depends on the local dataset $\mathcal{D}_m$, i.e.,

$$f_m(z) = \frac{1}{|\mathcal{D}_i|}\sum_{(x_j,y_j)\in\mathcal{D}_i} \ell(z; x_j, y_j),$$

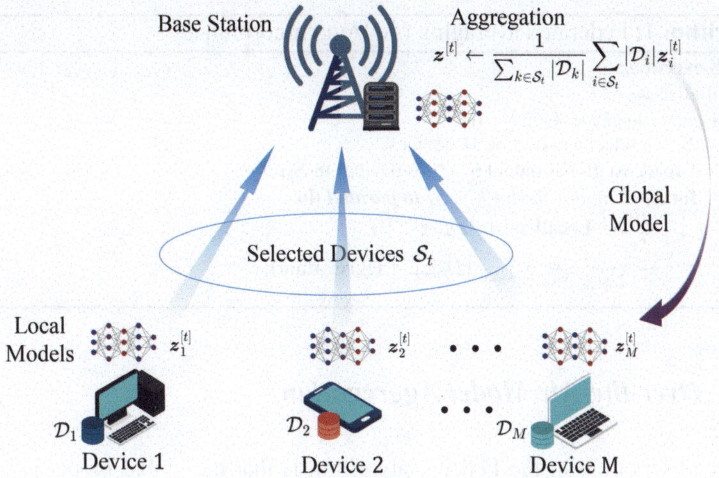

Fig. 2.1 Federated first-order optimization algorithm over wireless networks

where ℓ is a sample-wise loss function with the input-output data pair being $(\boldsymbol{x}_j, y_j)$, $\mathcal{D}_i$ is the local dataset at device i, and $|\mathcal{D}_i|$ denotes the number of data points in dataset $\mathcal{D}_i$.

2.2.2 *Federated Averaging Algorithm*

To improve the communication efficiency between edge devices and the BS for global model aggregation, the FedAvg algorithm proposed [10] is adopted. To be specific, at the t-th round, the following steps are iteratively performed.

1. The BS selects a subset of edge devices $\mathcal{S}_t \subseteq \{1, \ldots, M\}$;
2. The BS sends the updated global model $z^{[t-1]}$ to the selected devices $\mathcal{S}_t$;
3. Each selected device $i \in \mathcal{S}_t$ runs a local update algorithm (e.g., stochastic gradient algorithm) based on its local dataset $\mathcal{D}_i$ and the global model $z^{[t-1]}$, whose output is the updated local model $z_i^{[t]}$;
4. The BS aggregates all the local updates $z_i^{[t]}$ with $i \in \mathcal{S}_t$, i.e., computing a weighted average of local models as the updated global model $z^{[t]}$.

For convenience, the FedAvg framework adopted in this chapter is summarized in Algorithm 1.

Algorithm 1: Federated averaging (FedAvg) algorithm

BS executes:
Initialize $\boldsymbol{w}_0$.
for *each round* $t = 1, 2, \cdots$ **do**
  $S_t \leftarrow$ select a subset of M devices;
  Broadcast global model $z^{[t-1]}$ to devices in $\mathcal{S}_t$.
  for *Each mobile device* $k \in S_t$ ***in parallel*** **do**
    $z_k^{[t]} \leftarrow \text{LocalUpdate}(\mathcal{D}_k, z^{[t-1]})$
  $z^{[t]} \leftarrow \frac{1}{\sum_{k \in S_t} |\mathcal{D}_k|} \sum_{k \in S_t} |\mathcal{D}_k| z_k^{[t]}$ **(aggregation)**

2.2.3 *Over-the-Air Model Aggregation*

The key observation in the FedAvg algorithm is that the global model is updated through computing the weighted average of locally computed updates at each selected device, which falls in the category of computing nomographic functions of distributed data. We shall present an AirComp approach for communication-efficient model aggregation in FEEL system.

Specifically, the target vector for aggregating local updates in the FedAvg algorithm is given by

$$z = \psi \left(\sum_{i \in S_t} \phi_i(z_i) \right), \tag{2.2}$$

where z_i is the updated local model at the i-th device, $\phi_i = |\mathcal{D}_i|$ is the pre-processing scalar at device i, $\psi = \frac{1}{\sum_{k \in S_t} |\mathcal{D}_k|}$ is the post-processing scalar at the BS, and $\mathcal{S}$ is the selected set of mobile devices. The symbol vector for each local model before pre-processing $\boldsymbol{s}_i := z_i \in \mathbb{C}^d$ is assumed to be normalized with unit variance, i.e., $\mathbb{E}(\boldsymbol{s}_i \boldsymbol{s}_i^{\mathsf{H}}) = \boldsymbol{I}$. At each time slot $j \in \{1, \cdots, d\}$, each device sends signal $s_i^{(j)} \in \mathbb{C}$ to the BS. We denote

$$g^{(j)} = \sum_{i \in S_t} \phi_i \left(s_i^{(j)} \right) \tag{2.3}$$

as the target function to be estimated through over-the-air computation at the j-th time slot.

To simplify the notation, we omit the time index by writing $g^{(j)}$ and $s_i^{(j)}$ as g and s_i, respectively. The received signal at the BS is given by

$$\boldsymbol{y} = \sum_{i \in \mathcal{S}} \boldsymbol{h}_i b_i s_i + \boldsymbol{n}, \tag{2.4}$$

where $b_i \in \mathbb{C}$ is the transmitter scalar, $\boldsymbol{h}_i \in \mathbb{C}^N$ is the channel vector between device i and the BS, and $\boldsymbol{n} \sim \mathcal{CN}(\boldsymbol{0}, \sigma^2\boldsymbol{I})$ is the noise vector. The transmit power constraint at device i is given by

$$\mathbb{E}(|b_i s_i|^2) = |b_i|^2 \leq P_0 \tag{2.5}$$

with $P_0 > 0$ being the maximum transmit power. The estimated value before post-processing at the BS is given by

$$\hat{g} = \frac{1}{\sqrt{\eta}}\boldsymbol{m}^{\mathsf{H}}\boldsymbol{y} = \frac{1}{\sqrt{\eta}}\boldsymbol{m}^{\mathsf{H}}\sum_{i \in \mathcal{S}_t} \boldsymbol{h}_i b_i s_i + \frac{\boldsymbol{m}^{\mathsf{H}}\boldsymbol{n}}{\sqrt{\eta}}, \tag{2.6}$$

where $\boldsymbol{m} \in \mathbb{C}^N$ is the receive beamforming vector and η is a normalizing factor. Each element of the target vector can thus be obtained as $\hat{z} = \psi(\hat{g})$ at the BS.

The distortion of $\hat{g}$ with respect to the target value g given in (2.3), which quantifies the recovery performance for global model aggregation in the FedAvg algorithm, is measured by the following mean-squared-error (MSE)

$$\begin{aligned} \mathsf{MSE}(\hat{g}, g) &= \mathbb{E}\left(|\hat{g} - g|^2\right) \\ &= \sum_{i \in \mathcal{S}} \left|\boldsymbol{m}^{\mathsf{H}}\boldsymbol{h}_i b_i/\sqrt{\eta} - \phi_i\right|^2 + \sigma^2\|\boldsymbol{m}\|^2/\eta. \end{aligned} \tag{2.7}$$

We set the transmit scalar as $b_i = \sqrt{\eta}\phi_i \frac{(\boldsymbol{m}^{\mathsf{H}}\boldsymbol{h}_i)^{\mathsf{H}}}{\|\boldsymbol{m}^{\mathsf{H}}\boldsymbol{h}_i\|^2}$ to force the first term to be zero. Due to the transmit power constraint (2.5) for transmit scalar b_i, we have

$$\eta = \min_{i \in \mathcal{S}_t} \frac{P_0\|\boldsymbol{m}^{\mathsf{H}}\boldsymbol{h}_i\|^2}{\phi_i^2}. \tag{2.8}$$

The MSE is thus given by

$$\mathsf{MSE}(\hat{g}, g; \mathcal{S}_t, \boldsymbol{m}) = \frac{\|\boldsymbol{m}\|^2\sigma^2}{\eta} = \frac{\sigma^2}{P_0}\max_{i \in \mathcal{S}_t} \phi_i^2 \frac{\|\boldsymbol{m}\|^2}{\|\boldsymbol{m}^{\mathsf{H}}\boldsymbol{h}_i\|^2}. \tag{2.9}$$

In the following subsection, we shall characterize the convergence of the first-order FEEL algorithm in terms of the MSE performance.

2.2.4 Convergence Analysis

To make the analysis trackable, we make the following widely-adopted assumptions.

Assumption 1 *The global loss* $f(\boldsymbol{x})$ *is lower bounded by* f_*, *i.e.,* $f(\boldsymbol{x}) \geq f_* > -\infty$.

Assumption 2 $f(\boldsymbol{x})$ *is* L*-smooth. Mathematically, for any* $\boldsymbol{x} \in \mathbb{R}^d$ *and* $\boldsymbol{y} \in \mathbb{R}^d$, *we have*

$$f(\boldsymbol{y}) \leq f(\boldsymbol{x}) + \langle \nabla f(\boldsymbol{x}), \boldsymbol{y} - \boldsymbol{x} \rangle + \frac{L}{2} \|\boldsymbol{y} - \boldsymbol{x}\|^2 . \tag{2.10}$$

Assumption 3 *For any data sample* $(\boldsymbol{x}_j, y_j)$ *and model parameter* z, *we have* $\|\ell(z; \boldsymbol{x}_j, y_j)\|^2 \leq \kappa$.

The convergence performance of the AirComp-assisted FL algorithm is presented in the following proposition. It is worth noting that we only consider one-step gradient descent with learning rate ς at edge devices.

Lemma 2.1 *Suppose that Assumptions 1–3 hold, by setting the learning rate* $0 < \varsigma \leq 1/L$, *we have*

$$\begin{aligned} \frac{1}{T}\sum_{t=0}^{T-1} \mathbb{E}\left[\left\|\nabla f\left(z^{[t]}\right)\right\|^2\right] &\leq \frac{2}{\varsigma T}\left(f\left(z^{[0]}\right) - f\left(z^{\star}\right)\right) \\ &+ \frac{2}{T}\sum_{t=0}^{T-1}\left(4\kappa\left(1 - \frac{|S_t|}{M}\right)^2 + \frac{d}{\varsigma^2 |S_t|^2}\mathrm{MSE}^{[t]}\right), \end{aligned} \tag{2.11}$$

where T *denotes the total number of communication rounds and*

$$\mathrm{MSE}^{[t]} = \sum_{i \in S_t} \left|(\boldsymbol{m}^{[t]})^{\mathsf{H}} \boldsymbol{h}_i^{[t]} b_i^{[t]} / \sqrt{\eta^{[t]}} - \phi_i\right|^2 + \sigma^2 \|\boldsymbol{m}^{[t]}\|^2 / \eta^{[t]}. \tag{2.12}$$

Proof Starting from the iteration equation, we have

$$\begin{aligned} &z^{[t+1]} \\ &= z^{[t]} - \varsigma \frac{1}{K} \sum_{k \in \mathcal{K}} \nabla F_k\left(z^{[t]}\right) + \boldsymbol{e}_{\mathrm{cmm}}^{[t]} + \boldsymbol{e}_{\mathrm{sel}}^{[t]} \\ &= z^{[t]} - \varsigma\left(\nabla f\left(z^{[t]}\right) - \boldsymbol{e}^{[t]}\right), \end{aligned} \tag{2.13}$$

where $e^{[t]} = \left(e_{\text{cmm}}^{[t]} + e_{\text{sel}}^{[t]}\right)/\varsigma$, and $e_{\text{cmm}}^{[t]}$ and $e_{\text{sel}}^{[t]}$ denote the errors caused by communication and device selection, respectively, given by

$$\boldsymbol{e}_{\text{cmm}}^{[t]} = \frac{1}{|\mathcal{S}_t|}\left(\hat{\boldsymbol{g}}^{[t]} - \boldsymbol{g}^{[t]}\right),$$

$$\boldsymbol{e}_{\text{sel}}^{[t]} = \varsigma\left(\nabla f\left(z^{[t]}\right) - \frac{1}{|\mathcal{S}_t|}\sum_{k\in\mathcal{S}_t}\nabla f_k\left(z^{[t]}\right)\right).$$

Due to the smoothness of f, we have

$$\begin{aligned}
&f\left(z^{[t+1]}\right) - f\left(z^{[t]}\right)\\
&\leq \varsigma\left(\frac{L\varsigma}{2} - 1\right)\left\|\nabla f\left(z^{[t]}\right)\right\|^2 + (1 - L\varsigma)\varsigma\left\langle\nabla f\left(z^{[t]}\right), \boldsymbol{e}^{[t]}\right\rangle + \frac{L\varsigma^2}{2}\left\|e^{[t]}\right\|^2\\
&\leq -\frac{\varsigma}{2}\left\|\nabla f\left(z^{[t]}\right)\right\|^2 + \frac{\varsigma}{2}\left\|e^{[t]}\right\|^2\\
&\leq -\frac{\varsigma}{2}\left\|\nabla f\left(z^{[t]}\right)\right\|^2 + \frac{1}{\varsigma}\left(\left\|e_{\text{cmm}}^{[t]}\right\|^2 + \left\|e_{\text{sel}}^{[t]}\right\|^2\right).
\end{aligned} \tag{2.14}$$

According to [15], we have

$$\left\|e_{\text{sel}}^{[t]}\right\|^2 \leq 4\varsigma^2\kappa\left(1 - \frac{|\mathcal{S}_t|}{M}\right)^2. \tag{2.15}$$

Therefore, we have

$$\begin{aligned}
&\mathbb{E}\left[f\left(z^{[t+1]}\right)\right] - \mathbb{E}\left[f\left(z^{[t]}\right)\right]\\
&\leq -\frac{\varsigma}{2}\mathbb{E}\left[\left\|\nabla f\left(z^{[t]}\right)\right\|^2\right] + 4\varsigma\kappa\left(1 - \frac{|\mathcal{S}_t|}{M}\right)^2 + \frac{1}{\varsigma\,|\mathcal{S}_t|^2}\mathbb{E}\left[\left\|\hat{\boldsymbol{g}}^{[t]} - \boldsymbol{g}^{[t]}\right\|^2\right].
\end{aligned} \tag{2.16}$$

By telescoping the above inequality for $t = \{1, 2, \ldots, T\}$, we complete the proof.

□

2.2.5 Problem Formulation

As can be observed from (2.11), the optimization of the number of selected devices is a crucial factor for enhancing the learning performance in FEEL systems. Moreover, it is imperative to minimize the aggregation error resulting from AirComp transmission to mitigate any significant decline in the predictive accuracy of the model. To address these two challenges, we propose a novel method for identifying the maximum number of selected devices while ensuring that the MSE requirement for over-the-air model aggregation is met. This proposed approach entails formulating a mixed combinatorial optimization problem, which optimizes the selection of devices while also considering the MSE constraints as follows

$$\begin{aligned} &\underset{\mathcal{S}_t, \boldsymbol{m}\in\mathbb{C}^N}{\text{maximize}} \quad |\mathcal{S}_t| \\ &\text{subject to} \quad \left(\max_{i\in\mathcal{S}_t} \phi_i^2 \frac{\|\boldsymbol{m}\|^2}{\|\boldsymbol{m}^{\mathsf{H}}\boldsymbol{h}_i\|^2}\right) \leq \gamma, \end{aligned} \tag{2.17}$$

where $\gamma > 0$ is the MSE requirement for model aggregation. It will be demonstrated that problem (2.17) can be solved through the maximization of feasible non-convex quadratic constraints. To facilitate the efficient design of algorithms, we shall present a methodology based on sparse representation for identifying the maximum number of selected devices. Subsequently, the non-convex quadratic constraints shall be redefined as affine constraints with an additional rank-one constraint resulting from utilizing the matrix lifting technique.

2.3 Sparse and Low-Rank Optimization for FEEL

To facilitate algorithm design, we first rewrite problem (2.17) as a mixed combinatorial optimization problem with non-convex quadratic constraints as provided in Proposition 2.1.

Proposition 2.1 *Problem* (2.17) *is equivalent to the following mixed combinatorial optimization problem:*

$$\begin{aligned} &\underset{\mathcal{S}_t, \boldsymbol{m}\in\mathbb{C}^N}{\textit{maximize}} \quad |\mathcal{S}_t| \\ &\textit{subject to} \quad \|\boldsymbol{m}\|^2 - \gamma_i \|\boldsymbol{m}^{\mathsf{H}}\boldsymbol{h}_i\|^2 \leq 0, i \in \mathcal{S}_t, \\ &\qquad\qquad\quad \|\boldsymbol{m}\|^2 \geq 1, \end{aligned} \tag{2.18}$$

where $\gamma_i = \gamma/\phi_i^2$. *That is, our target becomes maximizing the number of feasible MSE constraints* $\|\boldsymbol{m}\|^2 - \gamma_i \|\boldsymbol{m}^{\mathsf{H}}\boldsymbol{h}_i\|^2 \leq 0$ *under the regularity condition* $\|\boldsymbol{m}\|^2 \geq 1$.

To maximize the number of feasible MSE constraints in problem (2.18), we can minimize the number of non-zero entries of $\boldsymbol{x}$, i.e.,

$$\begin{aligned} \underset{\boldsymbol{x}\in\mathbb{R}_+^M,\boldsymbol{m}\in\mathbb{C}^N}{\text{minimize}} \quad & \|\boldsymbol{x}\|_0 \\ \text{subject to} \quad & \|\boldsymbol{m}\|^2 - \gamma_i \|\boldsymbol{m}^{\mathsf{H}}\boldsymbol{h}_i\|^2 \le x_i, \forall i, \\ & \|\boldsymbol{m}\|^2 \ge 1. \end{aligned} \tag{2.19}$$

The sparsity structure of $\boldsymbol{x}$ indicates the feasibility of each mobile device. If $x_i = 0$, the i-th edge device can be selected while satisfying the MSE requirement.

However, both the MSE constraints and the regularity condition in problem (2.19) are non-convex quadratic constraints. To addressed this non-convexity issue, a natural way is to utilize the matrix lifting technique. Specifically, by lifting vector $\boldsymbol{m}$ as the positive semidefinite (PSD) matrix, i.e., $\boldsymbol{M} = \boldsymbol{m}\boldsymbol{m}^{\mathsf{H}}$ with $\text{rank}(\boldsymbol{M}) = 1$, problem (2.19) can be reformulated as the following sparse and low-rank optimization problem

$$\begin{aligned} \mathscr{P}: \underset{\boldsymbol{x}\in\mathbb{R}_+^M,\boldsymbol{M}\in\mathbb{C}^{N\times N}}{\text{minimize}} \quad & \|\boldsymbol{x}\|_0 \\ \text{subject to} \quad & \text{Tr}(\boldsymbol{M}) - \gamma_i \boldsymbol{h}_i^{\mathsf{H}}\boldsymbol{M}\boldsymbol{h}_i \le x_i, \forall i, \\ & \boldsymbol{M} \succeq \boldsymbol{0}, \text{Tr}(\boldsymbol{M}) \ge 1, \\ & \text{rank}(\boldsymbol{M}) = 1. \end{aligned} \tag{2.20}$$

Although problem $\mathscr{P}$ is still non-convex, we shall demonstrate its algorithmic advantages by developing efficient algorithms. Specifically, in the remaining part of this chapter, we shall first develop a parameter-free difference-of-convex (DC) approach to enhance sparsity, thereby maximizing the number of selected devices. Subsequently, we shall propose a novel DC approach to guarantee the exact rank-one constraint.

2.4 DC Representation for Sparse and Low-Rank Functions

Our goal is to introduce a unified framework for representing DC functions that can be applied to address the challenging problem of optimizing sparse and low-rank objectives in the context of FEEL with device selection. More precisely, our proposed approach leverages a novel DC representation for the ℓ_0-norm to induce sparsity and enable a structured method for device selection. Subsequently, we tackle a series of feasibility detection problems, involving non-convex quadratic constraints, to find the maximum number of devices that can participate in each communication round.

2.4.1 DC Representation for Sparse Function

Before introducing the DC representation for the ℓ_0-norm, we first give the definition of Ky Fan k-norm.

Definition 2.1 Ky Fan k-norm [3]: The Ky Fan k-norm of vector $\boldsymbol{x} \in \mathbb{C}^M$ is a convex function of $\boldsymbol{x}$ and is given by the sum of largest-k absolute values, i.e.,

$$|||\boldsymbol{x}|||_k = \sum_{i=1}^{M} |x_{\pi(i)}|, \tag{2.21}$$

where π is a permutation of $\{1, \cdots, M\}$ and $|x_{\pi(1)}| \geq \cdots \geq |x_{\pi(M)}|$.

If the ℓ_0-norm is less than k, its ℓ_1-norm is equal to its Ky Fan k-norm. Based on this fact, the ℓ_0-norm can be represented by the difference between ℓ_1-norm and Ky Fan k-norm [6], i.e.,

$$\|\boldsymbol{x}\|_0 = \min\{k : \|\boldsymbol{x}\|_1 - |||\boldsymbol{x}|||_k = 0, 0 \leq k \leq M\}. \tag{2.22}$$

2.4.2 DC Representation for Low-Rank Constraint

For the PSD matrix $\boldsymbol{M} \in \mathbb{C}^{N\times N}$, the rank-one constraint can be equivalently rewritten as

$$\sigma_i(\boldsymbol{M}) = 0, \forall i = 2, \cdots, N, \tag{2.23}$$

where $\sigma_i(\boldsymbol{M})$ is the i-th largest singular value of matrix $\boldsymbol{M}$. Note that the trace norm and spectral norm are, respectively, given by

$$\mathrm{Tr}(\boldsymbol{M}) = \sum_{i=1}^{N} \sigma_i(\boldsymbol{M}) \text{ and } \|\boldsymbol{M}\|_2 = \sigma_1(\boldsymbol{M}). \tag{2.24}$$

Therefore, we have the following proposition.

Proposition 2.2 *For PSD matrix $\boldsymbol{M}$ and $Tr(\boldsymbol{M}) > 0$, we have*

$$rank(\boldsymbol{M}) = 1 \Leftrightarrow Tr(\boldsymbol{M}) - \|\boldsymbol{M}\|_2 = 0. \tag{2.25}$$

2.4.3 A Unified DC Representation Framework

The DC representation framework aims to establish a sparsity pattern for vector $\boldsymbol{x}$ as its initial step. This pattern is essential in determining the priority order of device selection. Subsequently, a sequence of feasibility detection problems are solved to identify the maximum number of selected devices while ensuring the fulfillment of the MSE requirement.

2.4.3.1 Step I: Sparsity Inducing

In the first step, we solve the following DC program for problem $\mathscr{P}$:

$$\begin{aligned}\mathscr{P}_{\text{S1}} : \underset{\boldsymbol{x},\boldsymbol{M}}{\text{minimize}} \quad & \|\boldsymbol{x}\|_1 - \|\boldsymbol{x}\|_k + \text{Tr}(\boldsymbol{M}) - \|\boldsymbol{M}\|_2 \\ \text{subject to} \quad & \text{Tr}(\boldsymbol{M}) - \gamma_i \boldsymbol{h}_i^{\mathsf{H}} \boldsymbol{M} \boldsymbol{h}_i \le x_i, \forall i = 1, \cdots, M \\ & \boldsymbol{M} \succeq \boldsymbol{0}, \quad \text{Tr}(\boldsymbol{M}) \ge 1, \boldsymbol{x} \succeq \boldsymbol{0}.\end{aligned} \tag{2.26}$$

By sequentially solving problem $\mathscr{P}_{\text{S1}}$, we can derive the sparse vector $\boldsymbol{x}^\star$ with an objective value of zero, by progressively increasing the value of k from 0 to M. It should be emphasized that in achieving an objective value of zero through this process, the rank one constraint of matrix $\boldsymbol{M}$ must also be satisfied, i.e., $\text{Tr}(\boldsymbol{M}) - \|\boldsymbol{M}\|_2 = 0$.

2.4.3.2 Step II: Feasibility Detection

In the first step, the returned solution of $\boldsymbol{x}$ serves to depict the difference between the MSE requirement and the attainable MSE value for all devices. Thus, we propose to prioritize the selection of device k if x_k has a small value. The entries of $\boldsymbol{x}$ can be ordered in a descending sequence as $x_{\pi(1)} \ge \cdots \ge x_{\pi(M)}$. We determine the minimal value of k by progressively increasing k from 1 to M, such that choosing all the devices within $\mathcal{S}^{[k]}$ is feasible, where the set $\mathcal{S}^{[k]}$ is designated as $\{\pi(k), \pi(k+1), \cdots, \pi(M)\}$.

In detail, if all devices in $\mathcal{S}^{[k]}$ can be selected, the following optimization problem

$$\begin{aligned}\text{find} \quad & \boldsymbol{m} \\ \text{subject to} \quad & \|\boldsymbol{m}\|^2 - \gamma_i \|\boldsymbol{m}^{\mathsf{H}} \boldsymbol{h}_i\|^2 \le 0, \forall i \in \mathcal{S}^{[k]} \\ & \|\boldsymbol{m}\|^2 \ge 1.\end{aligned} \tag{2.27}$$

should be feasible. It can be equivalently reformulated as

$$\begin{aligned} &\text{find} \ \ \boldsymbol{M} \\ &\text{subject to} \ \ \mathrm{Tr}(\boldsymbol{M}) - \gamma_i \boldsymbol{h}_i^{\mathsf{H}} \boldsymbol{M} \boldsymbol{h}_i \leq 0, \forall i \in \mathcal{S}^{[k]} \\ &\qquad\qquad \boldsymbol{M} \succeq \boldsymbol{0}, \mathrm{Tr}(\boldsymbol{M}) \geq 1, \mathrm{rank}(\boldsymbol{M}) = 1 \end{aligned} \tag{2.28}$$

using the matrix lifting technique. To guarantee the feasibility of the fixed-rank constraint for accurately detecting the feasibility of MSE constraints, we propose the following DC approach by minimizing the difference between trace norm and spectral norm:

$$\begin{aligned} \mathscr{P}_{\mathrm{S2}} : \underset{\boldsymbol{M}}{\text{minimize}} \ \ & \mathrm{Tr}(\boldsymbol{M}) - \|\boldsymbol{M}\|_2 \\ \text{subject to} \ \ & \mathrm{Tr}(\boldsymbol{M}) - \gamma_i \boldsymbol{h}_i^{\mathsf{H}} \boldsymbol{M} \boldsymbol{h}_i \leq 0, \forall i \in \mathcal{S}^{[k]} \\ & \boldsymbol{M} \succeq \boldsymbol{0}, \quad \mathrm{Tr}(\boldsymbol{M}) \geq 1. \end{aligned} \tag{2.29}$$

When the objective value of problem $\mathscr{P}_{\mathrm{S2}}$ equals to zero for a given set $\mathcal{S}^{[k]}$, we conclude that all devices in $\mathcal{S}^{[k]}$ can be selected while satisfying the MSE requirement, i.e., problem (2.27) is feasible for $\mathcal{S}^{[k]}$. Note that the solution $\boldsymbol{M}^*$ shall be an exact rank-one matrix and a feasible receiver beamforming vector $\boldsymbol{m}$ can be obtained through Cholesky decomposition $\boldsymbol{M}^* = \boldsymbol{m}\boldsymbol{m}^{\mathsf{H}}$.

The DC representation framework for solving the sparse and low-rank optimization problem in FEEL is presented in Algorithm 2.

Algorithm 2: DC representation framework for solving problem $\mathscr{P}$ in FEEL with device selection

---

Step 1: Sparsity inducing
$k \leftarrow 0$
while *Objective value of* $\mathscr{P}_{S1}$ *is not zero* **do**
  Obtain solution $\boldsymbol{x}$ by solving the DC program $\mathscr{P}_{\mathrm{S1}}$
  $k \leftarrow k + 1$

Step 2: Feasibility detection
Order $\boldsymbol{x}$ in descending order as $x_{\pi(1)} \geq \cdots \geq x_{\pi(M)}$
$k \leftarrow 1$
while *Objective value of* $\mathscr{P}_{S2}$ *is not zero* **do**
  $\mathcal{S}^{[k]} \leftarrow \{\pi(k), \pi(k+1), \cdots, \pi(M)\}$
  Obtain solution $\boldsymbol{M}$ by solving the DC program $\mathscr{P}_{\mathrm{S2}}$
  $k \leftarrow k + 1$

Output: $\boldsymbol{m}$ through Cholesky decomposition $\boldsymbol{M} = \boldsymbol{m}\boldsymbol{m}^{\mathsf{H}}$, and the set of selected devices $\mathcal{S}^{[k]} = \{\pi(k), \pi(k+1), \cdots, \pi(M)\}$

2.5 DC Algorithm

2.5.1 *Difference-of-Strongly-Convex-Functions Representation*

We can equivalently rewrite problem $\mathscr{P}_{\mathrm{S1}}$ as

$$\underset{x,M}{\text{minimize}} \ f_1 = \|x\|_1 - |||x|||_k + \mathrm{Tr}(M) - \|M\|_2 + I_{C_1}(x, M), \tag{2.30}$$

and problem $\mathscr{P}_{\mathrm{S2}}$ as

$$\underset{M}{\text{minimize}} \ f_2 = \mathrm{Tr}(M) - \|M\|_2 + I_{C_2}(M), \tag{2.31}$$

respectively. Here, C_1 and C_2 are positive semidefinite cones that integrate the constraints of problem $\mathscr{P}_{\mathrm{S1}}$ and problem $\mathscr{P}_{\mathrm{S2}}$, and the indicator function is defined as

$$I_{C_1}(x, M) = \begin{cases} 0, & (x, M) \in C_1 \\ +\infty, & \text{otherwise} \end{cases}. \tag{2.32}$$

In order to establish the convergence result of the DC algorithm, we rewrite the DC functions f_1 and f_2 as the difference of *strongly* convex functions, i.e., $f_1 = g_1 - h_1$ and $f_2 = g_2 - h_2$, where

$$g_1 = \|x\|_1 + \mathrm{Tr}(M) + I_{C_1}(x, M) + \frac{\alpha}{2}(\|x\|_F^2 + \|M\|_F^2), \tag{2.33}$$

$$h_1 = |||x|||_k + \|M\|_2 + \frac{\alpha}{2}(\|x\|_F^2 + \|M\|_F^2), \tag{2.34}$$

$$g_2 = \mathrm{Tr}(M) + I_{C_2}(M) + \frac{\alpha}{2}\|M\|_F^2, \tag{2.35}$$

$$h_2 = \|M\|_2 + \frac{\alpha}{2}\|M\|_F^2. \tag{2.36}$$

By adding quadratic terms, g_1, g_2, h_1, and h_2 are all α-strongly convex functions. Subsequently, problem (2.30) and problem (2.31) admit the uniform structure of minimizing the difference of two strongly convex functions, i.e.,

$$\underset{X \in \mathbb{C}^{m \times n}}{\text{minimize}} \ f(X) = g(X) - h(X). \tag{2.37}$$

The DC algorithm is given by constructing a solution sequence.

2.5.2 DC Algorithm for Sparse and Low-Rank Optimization

The t-th iteration of the simplified DC algorithm is to solve the convex approximation of the problem by linearizing its concave part, i.e.,

$$\boldsymbol{X}^{[t+1]} = \arg \inf_{\boldsymbol{X} \in \mathcal{X}} g(\boldsymbol{X}) - [h(\boldsymbol{X}^{[t]}) + \langle \boldsymbol{X} - \boldsymbol{X}^{[t]}, \boldsymbol{Y}^{[t]} \rangle], \tag{2.38}$$

where

$$\boldsymbol{Y}^{[t]} \in \partial_{\boldsymbol{X}^{[t]}} h, \tag{2.39}$$

$\partial_{\boldsymbol{X}^{[t]}} h$ is the subgradient of h with respect to $\boldsymbol{X}$ at $\boldsymbol{X}^{[t]}$.

Therefore, the iterations of $\boldsymbol{x}^{[t]}$ and $\boldsymbol{M}^{[t]}$ of the DC algorithm for problem $\mathscr{P}_{\mathrm{S1}}$ are constructed as the solution to the following convex optimization problem

$$\begin{aligned} \underset{\boldsymbol{x}, \boldsymbol{M}}{\text{minimize}} \quad & g_1 - \langle \partial_{\boldsymbol{x}^{[t-1]}} h_1, \boldsymbol{x} \rangle - \langle \partial_{\boldsymbol{M}^{[t-1]}} h_1, \boldsymbol{M} \rangle \\ \text{subject to} \quad & \mathrm{Tr}(\boldsymbol{M}) - \gamma_i \boldsymbol{h}_i^{\mathsf{H}} \boldsymbol{M} \boldsymbol{h}_i \le x_i, \forall i = 1, \cdots, M, \\ & \boldsymbol{M} \succeq \boldsymbol{0}, \quad \mathrm{Tr}(\boldsymbol{M}) \ge 1, \boldsymbol{x} \succeq \boldsymbol{0}. \end{aligned} \tag{2.40}$$

The iteration of $\boldsymbol{M}^{[t]}$ for problem $\mathscr{P}_{\mathrm{S2}}$ is given by the solution to the following optimization problem

$$\begin{aligned} \underset{\boldsymbol{M}}{\text{minimize}} \quad & g_2 - \langle \partial_{\boldsymbol{M}^{[t-1]}} h_2, \boldsymbol{M} \rangle \\ \text{subject to} \quad & \mathrm{Tr}(\boldsymbol{M}) - \gamma_i \boldsymbol{h}_i^{\mathsf{H}} \boldsymbol{M} \boldsymbol{h}_i \le 0, \forall i \in \mathcal{S}^{[k]}, \\ & \boldsymbol{M} \succeq \boldsymbol{0}, \quad \mathrm{Tr}(\boldsymbol{M}) \ge 1. \end{aligned} \tag{2.41}$$

The subgradient of h_1 and h_2 are given by

$$\partial_{\boldsymbol{x}} h_1 = \partial |||\boldsymbol{x}|||_k + \alpha \boldsymbol{x}, \tag{2.42}$$

$$\partial_{\boldsymbol{M}} h_1 = \partial_{\boldsymbol{M}} h_2 = \partial \|\boldsymbol{M}\|_2 + \alpha \boldsymbol{M}. \tag{2.43}$$

The subgradient of $|||\boldsymbol{x}|||_k$ can be computed by Gotoh et al. [6]

$$i\text{-th entry of } \partial |||\boldsymbol{x}|||_k = \begin{cases} \mathrm{sign}(x_i), & |x_i| \ge |x_{(k)}| \\ 0, & |x_i| < |x_{(k)}| \end{cases}. \tag{2.44}$$

The subgradient of $\|\boldsymbol{M}\|_2$ is given by the following proposition.

Proposition 2.3 *The subgradient of* $\|\boldsymbol{M}\|_2$ *can be computed as* $\boldsymbol{v}_1\boldsymbol{v}_1^{\mathsf{H}}$, *where* $\boldsymbol{v}_1 \in \mathbb{C}^N$ *is the eigenvector of the largest eigenvalue* $\sigma_1(\boldsymbol{M})$.

2.6 Simulation Results

We next conduct simulations to compare the performance of the proposed DC method with that of the state-of-the-art approaches for FEEL with device selection. The channel coefficient vectors $\boldsymbol{h}_i$'s between the BS and each device follow the i.i.d. complex normal distribution, i.e., $\boldsymbol{h}_i \sim \mathcal{CN}(\boldsymbol{0}, \boldsymbol{I})$. The average transmit signal-to-noise-ratio (SNR) P_0/σ^2 is set to 20 dB. We assume that all devices have the same number of data samples, i.e., $|\mathcal{D}_1| = \cdots = |\mathcal{D}_M|$, for which the pre-processing and post-processing pair can be chosen as $\phi_i = 1$, $\psi = 1/|\mathcal{S}|$, respectively.

To show the performance of the proposed DC approach for device selection in FEEL, we train a support vector machine (SVM) classifier on CIFAR-10 dataset [7] with a 6-antenna BS and 20 mobile devices. The CIFAR-10 dataset, widely used for images classification, contains 10 different classes of objects. We set the benchmark as the scenario where all devices are selected and all local updates are aggregated without aggregation error. Under the presented two-step framework and ordering rule in Algorithm 2, we compare the proposed DC Algorithm 2 for device selection with the following state-of-the-art approaches:

- ℓ_1 **+ SDR**: The ℓ_1-norm minimization is adopted to induce the sparsity of $\boldsymbol{x}$ in Step 1, and the nonconvex quadratic constraints are addressed with SDR in Step 1 and Step 2.
- **Reweighted** ℓ_2 **+ SDR**: We take the smoothed ℓ_p-norm for sparsity inducing of $\boldsymbol{x}$ in Step 1, which is solved by the reweighted ℓ_2-minimization algorithm. The SDR approach is used to address the nonconvex quadratic program in Step 1 and Step 2.

As shown in Figs. 2.2 and 2.3, respectively, the simulation results demonstrate that the proposed DC approach achieves lower training loss and higher prediction accuracy than the baselines. This is because, although the semidefinite relaxation (SDR) technique convexifies the nonconvex quadratic constraints as a linear constraint via dropping the rank-one constraint in the lifting problem, the performance degenerates with large number of antennas due to its weak capability of inducing low-rank structures [2]. Additionally, our proposed AirComp-assisted FedAvg algorithm achieves the same performance to the benchmark which is independent of the channel fading and receiver noise.

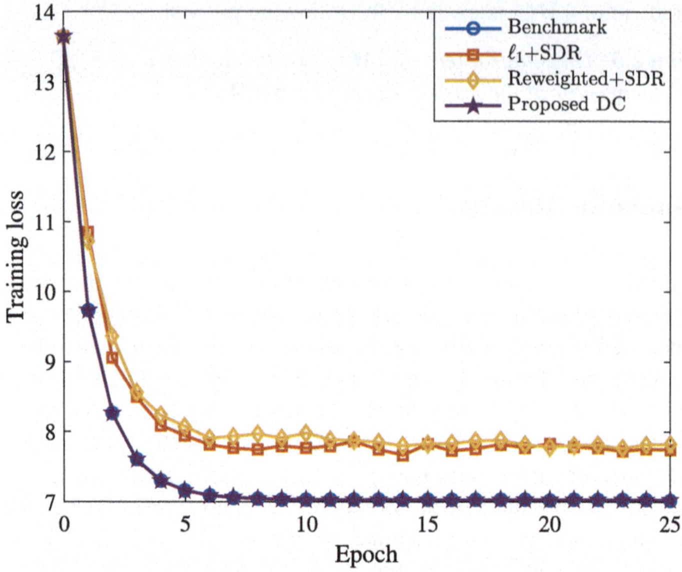

Fig. 2.2 Convergence of different device selection algorithms for FedAvg

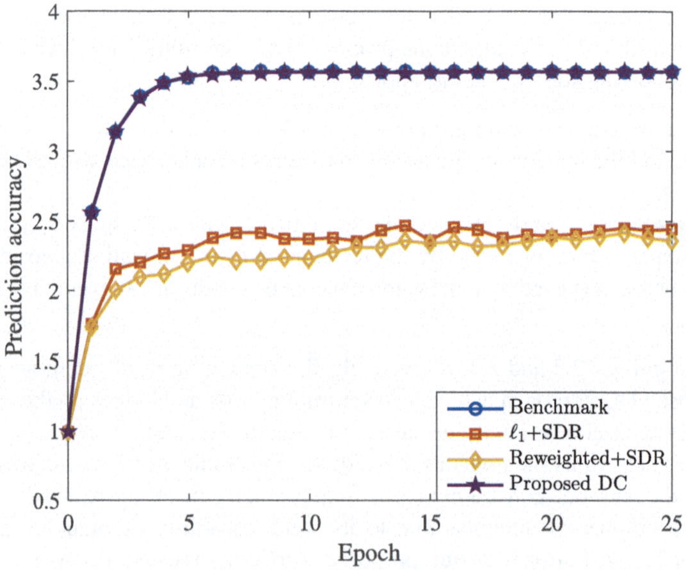

Fig. 2.3 The relationship between communication rounds and test accuracy

2.7 Summary

In this chapter, we presented a novel fast global model aggregation approach for federated first order method based on the principles of AirComp. To improve the learning performance for FEEL, we developed a novel sparse and low-rank modeling approach to maximize the selected devices with the MSE requirements for model aggregation. To achieve sparsity and low-rankness, we proposed a unified DC representation framework bolstered by our introduced DC algorithm through successive convex relaxation. Evidenced through simulation results, the proposed techniques exhibited remarkable performance when compared with the state-of-the-art algorithms.

References

1. M. M. Amiri and D. Gündüz. Machine learning at the wireless edge: Distributed stochastic gradient descent over-the-air. *IEEE Trans. Signal Process.*, 68:2155–2169, Mar. 2020.
2. L. Chen, X. Qin, and G. Wei. A uniform-forcing transceiver design for over-the-air function computation. *IEEE Wireless Commun. Lett.*, 7(6):942–945, 2018.
3. K. Fan. Maximum properties and inequalities for the eigenvalues of completely continuous operators. *Proceedings of the National Academy of Sciences*, 37(11):760–766, 1951.
4. W. Fang, Z. Yu, Y. Jiang, Y. Shi, C. N. Jones, and Y. Zhou. Communication-efficient stochastic zeroth-order optimization for federated learning. *IEEE Transactions on Signal Processing*, 70:5058–5073, 2022.
5. W. Fang, Y. Zou, H. Zhu, Y. Shi, and Y. Zhou. Optimal receive beamforming for over-the-air computation. In *2021 IEEE 22nd International Workshop on Signal Processing Advances in Wireless Communications (SPAWC)*, pages 61–65. IEEE, 2021.
6. J.-y. Gotoh, A. Takeda, and K. Tono. Dc formulations and algorithms for sparse optimization problems. *Mathematical Programming*, 169:141–176, 2018.
7. A. Krizhevsky, G. Hinton, et al. Learning multiple layers of features from tiny images. 2009.
8. K. B. Letaief, Y. Shi, J. Lu, and J. Lu. Edge artificial intelligence for 6G: Vision, enabling technologies, and applications. *IEEE J. Sel. Areas Commun.*, pages 1–1, 2021, doi:10.1109/JSAC.2021.3126076.
9. Q. Lu and L. Lian. Decentralized complete dictionary learning via ℓ^4-norm maximization, 2022.
10. B. McMahan, E. Moore, D. Ramage, S. Hampson, and B. A. y. Arcas. Communication-Efficient Learning of Deep Networks from Decentralized Data. In *Proc. Int. Conf. Artif. Intell. Stat. (AISTATS)*, 2017.
11. Y. Shi, K. Yang, Z. Yang, and Y. Zhou. *Mobile Edge Artificial Intelligence: Opportunities and Challenges*. Elsevier, 2021.
12. Y. Shi, Y. Zhou, D. Wen, Y. Wu, C. Jiang, and K. B. Letaief. Task-oriented communications for 6G: Vision, principles, and technologies. *IEEE Wireless Commun.*, 30(3):78–85, Mar. 2023.
13. J. Wang, Q. Liu, H. Liang, G. Joshi, and H. V. Poor. A novel framework for the analysis and design of heterogeneous federated learning. *IEEE Trans. Signal Process.*, 69:5234–5249, 2021.
14. S. Wang, S. Hosseinalipour, V. Aggarwal, C. G. Brinton, D. J. Love, W. Su, and M. Chiang. Towards cooperative federated learning over heterogeneous edge/fog networks. *IEEE Communications Magazine*, 2023.

15. Z. Wang, J. Qiu, Y. Zhou, Y. Shi, L. Fu, W. Chen, and K. B. Letaief. Federated learning via intelligent reflecting surface. *IEEE Transactions on Wireless Communications*, 21(2):808–822, 2021.
16. K. Yang, T. Jiang, Y. Shi, and Z. Ding. Federated learning via over-the-air computation. *IEEE Trans. Wirel. Commun.*, 19(3):2022–2035, 2020.
17. X. Zhang, M. Hong, S. Dhople, W. Yin, and Y. Liu. FedPD: A federated learning framework with adaptivity to non-IID data. *IEEE Trans. Signal Process.*, 69:6055–6070, Oct. 2021.
18. Y. Zhou, Y. Shi, H. Zhou, J. Wang, L. Fu, and Y. Yang. Toward scalable wireless federated learning: Challenges and solutions. *IEEE Internet Things Mag.*, 6(4):10–16, Apr. 2023.
19. G. Zhu, J. Xu, K. Huang, and S. Cui. Over-the-air computing for wireless data aggregation in massive IoT. *IEEE Wirel. Commun.*, 28(4):57–65, Aug. 2021.

Chapter 3
Second-Order Algorithm for Federated Edge Learning

3.1 Overview

Extensive research has been undertaken in the domain of federated optimization to address the communication predicament. A predominant technique involves the reduction of the aggregate communication rounds. A plethora of gradient descent-based approaches have demonstrated that augmenting the local computation can yield a remarkable reduction in the total count of communication rounds [19, 22, 27, 32]. For example, FedAvg algorithm [19] performs multiple local updates to accelerate the training process. Motivated by FedAvg, a series of research focusing on the local model updates has been conducted in [17, 27]. In [32], an accelerated version of FedAvg was developed by parallelizing a generalized version of accelerated stochastic gradient descent (SGD). In [22], the authors propose a new federated optimization algorithm based on operator splitting to achieve better convergence rates and more accurate solutions. Nevertheless, these existing approaches are governed by the linear convergence in the best case. As a result, the total number of iteration rounds required to achieve the desired accuracy is relatively large [1]. Therefore, the second-order methods (e.g., Newton-type methods) become attractive in such a wireless environment due to their fast local quadratic convergence rate. However, the construction of the canonical Newton update requires both the Hessian and gradient information, where the distributed situation in FEEL makes gathering Hessian information a severe communication overhead. To this end, the second-order federated optimization algorithms have been investigated to resolve this issue [1, 5, 6, 25, 33]. The fast convergence rate with efficient communication makes the application of these second-order algorithms a great benefit to FEEL.

In spite of the potential in the application of second-order algorithms to reduce the total communication rounds and improve the communication efficiency, the transmission of FEEL model parameters through wireless channels still confronts great challenges as wireless channels are always noisy with limited resources and

Y. Zhou et al., *Federated Edge Learning*, Wireless Networks,
https://doi.org/10.1007/978-3-031-96649-1_3

high latency [15, 16, 30]. Given to these issue, this chapter will take AirComp [8, 9, 29] to support the implementation of the proposed second-order FEEL algorithm over the wireless network [31]. We theoretically analyze the convergence behaviors of our proposed optimization algorithm with the presence of data heterogeneity, device selection, and channel noise. The analysis shows that our algorithm keeps a linear-quadratic convergence rate and outperforms first-order methods. To minimize the error gap and enhance performance, we formulate the problem as a combinatorial non-convex problem and propose a system optimization approach that combines Gibbs Sampling and the DC algorithm.

3.2 Learning Model and Algorithm

3.2.1 Learning Model

As shown in Fig. 3.1, there are m single-antenna edge devices and an edge server equipped with k antennas to collaboratively complete a learning task. We denote $\mathcal{D}$ as the entire sample set used in the FEEL task. Each device $i \in \mathcal{S}$ stores a sample set $\mathcal{D}_i = \left\{z_{i,j} := \left(\boldsymbol{u}_{i,j}, v_{i,j}\right)\right\}$, where $\mathcal{S}$ denotes the index set of edge devices, $\boldsymbol{u}_{i,j}$ is the feature vector, and $v_{i,j}$ is the corresponding label. The data set of all edge devices is denoted as $\mathcal{D} = \bigcup_{i=1}^{m} \mathcal{D}_i$ with $|\mathcal{D}| = n$.

The local loss function of the i-th device is defined as

$$F_i(\boldsymbol{w}) = \frac{1}{|\mathcal{D}_i|} \sum_{z_{i,j} \in \mathcal{D}_i} f\left(\boldsymbol{w}, z_{i,j}\right) + \frac{\gamma}{2} \|\boldsymbol{w}\|_2^2, \tag{3.1}$$

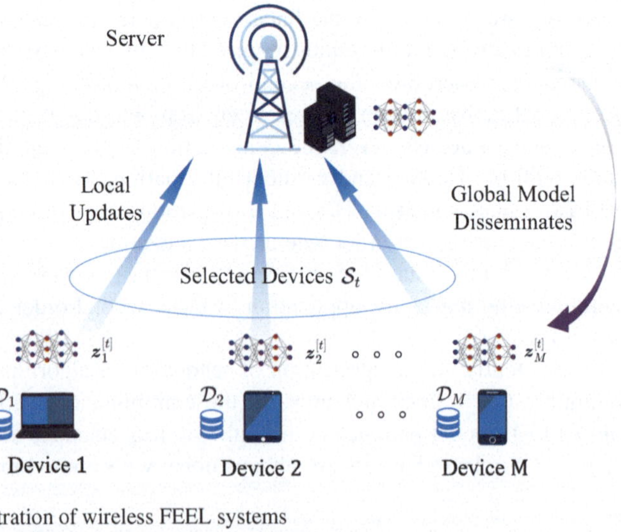

Fig. 3.1 Illustration of wireless FEEL systems

where the first term is the average of $f\left(\boldsymbol{w}, z_{i,j}\right)$, where $\boldsymbol{w} \in \mathbb{R}^d$ is the model parameter vector and function f is used to measure the prediction error of $\boldsymbol{w}$. The second term is added for regularization with γ being the weighting parameter. FEEL aims to train a suitable model at the edge server by aggregating the local model updates from multiple edge devices, on which the distributed models are trained based on the local datasets. Specifically, the edge server optimizes the following global loss function

$$F(\boldsymbol{w}) = \frac{1}{n} \sum_{i=1}^{m} |\mathcal{D}_i| F_i(\boldsymbol{w}) . \tag{3.2}$$

3.2.2 Federated Second-Order Optimization Algorithm

This chapter considers a second-order optimization algorithm with a fast convergence rate to significantly reduce the number of communication rounds. The descent direction vector of canonical Newton's method [21] is given by

$$\boldsymbol{p} = \left(\nabla^2 F(\boldsymbol{w})\right)^{-1} \nabla F(\boldsymbol{w}) . \tag{3.3}$$

The canonical Newton's method can achieve a locally quadratic convergence rate so that its total number of iteration rounds needed to complete the learning task is much smaller than that of the first-order optimization algorithms. However, in the distributed scenario, the computation of $\nabla^2 F(\boldsymbol{w}) = \frac{1}{m}\sum_{i=1}^{m} \nabla^2 F_i(\boldsymbol{w})$ requires the aggregation of the local Hessian $\nabla^2 F_i(\boldsymbol{w})$. The transmission of such $d \times d$ matrices inevitably brings huge communication overheads in each communication round. To solve this issue, numerous second-order distributed ML algorithms have been proposed, such as DANE [23], DISCO [33], GIANT [25], DINGO [5], and DINO [6]. These methods approximate Hessian information in various forms to avoid the direct transmission of Hessian matrices and approach the performance of canonical Newton's method. However, at least two communication rounds per iteration are required, including the aggregation of local gradients and second-order descent directions. Different from these second-order algorithms that require the aggregation of local gradients $\nabla F_i(\boldsymbol{w})$ to compute the global gradient $\nabla F(\boldsymbol{w}) = \frac{1}{m}\sum_{i=1}^{m} \nabla F_i(\boldsymbol{w})$, a recently proposed COMRADE [11] method cuts down the communication overhead of the aggregation. By this means, the number of communication rounds required per iteration is reduced to one, thereby further improving the communication efficiency. Motivated by this, we leverage the local Newton step aggregation as in [11] to achieve a fast convergence rate by reducing the number of communication rounds. The product of the inversion of the local Hessian matrix $\left(\nabla^2 F_i(\boldsymbol{w})\right)^{-1}$ and the local gradient $\nabla F_i(\boldsymbol{w})$ is used to serve as the local descent direction vector $\boldsymbol{p}_i = \left(\nabla^2 F_i(\boldsymbol{w})\right)^{-1} \nabla F_i(\boldsymbol{w})$ for model aggregation.

In this way, with the preserved convergence behavior of Newton's method, only one aggregation of the d-dimensional local descent direction vectors is carried out in each iteration. To be specific, at the t-th iteration, the procedure of our proposed method is summarized as follows:

(1) **Device Selection**: The server decides the set of devices, denoted as $\mathcal{S}_t$, to participate in this iteration.
(2) **Global Model Broadcast**: The server disseminates the current global model parameter vector $\boldsymbol{w}_t$ to the selected devices through the wireless channel.
(3) **Local Model Update**: After the i-th device receives global model parameter vector $\boldsymbol{w}_t$, it first computes the local gradient based on local data samples:

$$g_{t,i} = \nabla F_i(\boldsymbol{w}_t) = \frac{1}{|\mathcal{D}_i|} \sum_{z_{i,j} \in \mathcal{D}_i} \nabla f\left(\boldsymbol{w}_t, z_{i,j}\right) + \gamma \boldsymbol{w}_t \,, \tag{3.4}$$

where the derivatives are taken with respect to the first argument. Afterwards, the i-th device calculates the local Hessian matrix according to local gradient and local data samples:

$$\boldsymbol{H}_{t,i} = \nabla^2 F_i(\boldsymbol{w}_t) = \frac{1}{|\mathcal{D}_i|} \sum_{z_{i,j} \in \mathcal{D}_i} \nabla^2 f\left(\boldsymbol{w}_t, z_{i,j}\right) + \gamma \boldsymbol{I}_d \,. \tag{3.5}$$

The i-th device then gets a local Newton descent direction vector from previous results:

$$\boldsymbol{p}_{t,i} = \boldsymbol{H}_{t,i}^{-1} g_{t,i} = \left(\nabla^2 F_i(\boldsymbol{w}_t)\right)^{-1} \nabla F_i(\boldsymbol{w}_t) \,. \tag{3.6}$$

In practice, this step involves the computation of Hessian matrix and its inverse operation. To reduce the computational complexity, we adopt the conjugate gradient method [21] to obtain an approximate local Newton descent direction vector. According to the analysis in [25], this approximate solution will not have a significant impact on the convergence behavior.
(4) **Model Aggregation**: The devices participating in the t-th iteration transmit local Newton descent direction vectors $\{\boldsymbol{p}_{t,i}\}$ to the server through the wireless channel, and the server aggregates them to obtain the global descent direction vector for this iteration:

$$\tilde{\boldsymbol{p}}_t = \frac{1}{\sum_{i \in \mathcal{S}_t} |\mathcal{D}_i|} \sum_{i \in \mathcal{S}_t} |\mathcal{D}_i| \, \boldsymbol{p}_{t,i} \,. \tag{3.7}$$

(5) **Global Model Update**: The server updates the model parameter vector $\boldsymbol{w}_t$ based on global descent direction vector $\tilde{\boldsymbol{p}}_t$ and learning rate α.

$$\boldsymbol{w}_{t+1} = \boldsymbol{w}_t - \alpha \tilde{\boldsymbol{p}}_t \,. \tag{3.8}$$

3.2.3 Communication Model

In this chapter, we consider a block fading channel. Each block is divided into d time slots, ensuring the transmission of one local descent direction vector. We adopt AirComp [20] to support the model uploading. Before transmission, $\boldsymbol{p}_{t,i}$ is first preprocessed and then encoded as $\boldsymbol{s}_{t,i} \in \mathbb{R}^d$ at the i-th device:

$$\boldsymbol{s}_{t,i} = \phi_i\left(\boldsymbol{p}_{t,i}\right) = \frac{|\mathcal{D}_i|\, \boldsymbol{p}_{t,i}}{\bar{p}_{t,i}}, \tag{3.9}$$

where $\bar{p}_{t,i} = |\mathcal{D}_i| \left\|\boldsymbol{p}_{t,i}\right\|$ is the product of the size of the local dataset and the magnitude of $\boldsymbol{p}_{t,i}$. In this way, the stationary of the information-bearing symbols $\{\boldsymbol{s}_{t,i}\}$ can be guaranteed. Hence, we have $\left\|\boldsymbol{s}_{t,i}\right\|^2 = 1$ and $\mathbb{E}\left(|\boldsymbol{s}_{t,i}[j]|^2\right) = \frac{1}{d}$, $\forall j \in d$, where $\boldsymbol{s}_{t,i}[j]$ denotes the j-th entry of $\boldsymbol{s}_{t,i}$. Thereafter, each entry of the transmitted signal sent by the i-th device is given by

$$\boldsymbol{x}_{t,i}[j] = b_{t,i}\boldsymbol{s}_{t,i}[j], \tag{3.10}$$

where $\boldsymbol{x}_{t,i}[j] \in \mathbb{R}$ and $\boldsymbol{s}_{t,i}[j] \in \mathbb{R}$ denote two representative entries of $\boldsymbol{x}_{t,i}$ and $\boldsymbol{s}_{t,i}$, respectively. $b_{t,i} \in \mathbb{R}$ is the transmit power control factor, and the power constraint for each device in the whole process is given by:

$$\mathbb{E}\left(|b_{t,i}\boldsymbol{s}_{t,i}[j]|^2\right) = b_{t,i}^2/d \leq P_0, \ \forall t, i, \tag{3.11}$$

where P_0 denotes the maximum transmit power of each device.

Let $\boldsymbol{h}_{t,i} \in \mathbb{C}^k$ be the channel coefficient vector between the i-th device and the server in the t-th block, which remains unchanged in each block but differs among blocks. Then, the received signal $\boldsymbol{y}_t \in \mathbb{C}^k$ at the server can be represented as

$$\boldsymbol{y}_t = \sum_{i \in \mathcal{S}_t} \boldsymbol{h}_{t,i}\boldsymbol{x}_{t,i}[j] + \boldsymbol{e}_t = \sum_{i \in \mathcal{S}_t} \tilde{\boldsymbol{h}}_{t,i} b_{t,i} |\mathcal{D}_i|\, \boldsymbol{p}_{t,i}[j] + \boldsymbol{e}_t, \tag{3.12}$$

where $\tilde{\boldsymbol{h}}_{t,i} = \frac{\boldsymbol{h}_{t,i}}{\bar{p}_{t,i}}$ is the effective channel coefficient introduced in [12], $\boldsymbol{e}_t \in \mathbb{C}^k$ denotes the additive white Gaussian noise (AWGN) vector with the power of σ^2. We define the signal-to-noise ratio (SNR) as P_0/σ^2.

After receiving $\boldsymbol{y}_t$, the edge server can obtain $\boldsymbol{r}_t[j] \in \mathbb{C}$ before post-processing:

$$\boldsymbol{r}_t[j] = \frac{1}{\sqrt{\eta_t}}\boldsymbol{a}_t^{\mathsf{H}}\boldsymbol{y}_t = \frac{1}{\sqrt{\eta_t}}\left(\boldsymbol{a}_t^{\mathsf{H}} \sum_{i \in \mathcal{S}_t} \tilde{\boldsymbol{h}}_{t,i} b_{t,i} |\mathcal{D}_i|\, \boldsymbol{p}_{t,i}[j] + \boldsymbol{a}_t^{\mathsf{H}}\boldsymbol{e}_t\right), \tag{3.13}$$

where $\boldsymbol{a}_t \in \mathbb{C}^k$ represents the receive beamforming vector and η_t is the scaling factor. For convenience, we use $\boldsymbol{H}_t = [\tilde{\boldsymbol{h}}_{t,1}, \ldots, \tilde{\boldsymbol{h}}_{t,|\mathcal{S}_t|}]$ to denote the effective channel coefficient matrix, $\boldsymbol{B}_t = \operatorname{diag}\left(b_{t,1}, \ldots, b_{t,|\mathcal{S}_t|}\right)$ to denote the power transmission matrix, $\boldsymbol{G}_t = [|\mathcal{D}_1|\, \boldsymbol{p}_{t,1}, \ldots, \left|\mathcal{D}_{|\mathcal{S}_t|}\right| \boldsymbol{p}_{t,|\mathcal{S}_t|}]^{\mathsf{T}}$ to denote the signal transmission matrix, and $\boldsymbol{E}_t = [\boldsymbol{e}_{t,1}, \ldots, \boldsymbol{e}_{t,d}]$ to denote the noise matrix. Thereby, the total estimated value vector $\boldsymbol{r}_t = [\boldsymbol{r}_t[1], \ldots, \boldsymbol{r}_t[d]]$ can be written as

$$\boldsymbol{r} = \frac{1}{\sqrt{\eta_t}} \left(\boldsymbol{a}^{\mathsf{H}} \boldsymbol{H}_t \boldsymbol{B}_t \boldsymbol{G}_t + \boldsymbol{a}_t^{\mathsf{H}} \boldsymbol{E}_t \right) . \tag{3.14}$$

To alleviate the influence of the distortion caused by noise and improve the performance of AirComp, each entry of $\boldsymbol{B}_t$ is set to

$$b_{t,i} = \sqrt{\eta_t} \frac{\left(\boldsymbol{a}_t^{\mathsf{H}} \tilde{\boldsymbol{h}}_{t,i} \right)^{\mathsf{H}}}{\left\| \boldsymbol{a}_t^{\mathsf{H}} \tilde{\boldsymbol{h}}_{t,i} \right\|^2} . \tag{3.15}$$

By substituting (3.15) into (3.14), we can get a simplified version of $\boldsymbol{r}_t$

$$\boldsymbol{r}_t = \sum_{i \in \mathcal{S}_t} |\mathcal{D}_i|\, \boldsymbol{p}_{t,i}^{\mathsf{T}} + \frac{1}{\sqrt{\eta_t}} \boldsymbol{a}_t^{\mathsf{H}} \boldsymbol{E}_t . \tag{3.16}$$

Finally, through the post-processing function of ψ, the server obtains the global descent direction vector $\hat{\boldsymbol{p}}_t$

$$\hat{\boldsymbol{p}}_t = \psi\left(\boldsymbol{r}_t\right) = \frac{1}{\sum_{i \in \mathcal{S}_t} |\mathcal{D}_i|} \boldsymbol{r}_t^{\mathsf{H}} = \tilde{\boldsymbol{p}}_t + \frac{1}{\left(\sum_{i \in \mathcal{S}_t} |\mathcal{D}_i|\right) \sqrt{\eta_t}} \left(\boldsymbol{a}_t^{\mathsf{H}} \boldsymbol{E}_t \right)^{\mathsf{H}} , \tag{3.17}$$

where $\tilde{\boldsymbol{p}}_{t,i}$ is the averaged local descent direction vector as defined in (3.7).

Based on the AirComp-based communication model and second-order optimization algorithm, we propose an over-the-air federated second-order optimization algorithm, as shown in Algorithm 3.

3.3 Convergence Analysis

A major challenge of convergence analysis is to tackle the distortion of the descent direction vector caused by channel noise, device selection, and the use of local Newton step. To address this issue, we study the impact of distortion with respect to these influencing factors. In particular, we exploit the idea of sketching to analyze the approximation of local gradients and Hessian matrices. To better elaborate our analysis, some preliminaries are firstly presented.

Algorithm 3: Over-the-air federated second-order algorithm

for *Each iteration t* **do**
- Edge server chooses devices participating in this iteration and stores them as $\mathcal{S}_t$.
- Edge server broadcasts the current model parameter vector $\boldsymbol{w}_t$ to all devices.
- **for** *Each participating device i* ***in parallel*** **do**
  - Compute local gradient $\boldsymbol{g}_{t,i} = \frac{1}{|\mathcal{D}_i|}\sum_{z_{i,j}\in\mathcal{D}_i} \nabla f\left(\boldsymbol{w}_t, z_{i,j}\right) + \gamma \boldsymbol{w}_t$.
  - Compute local Hessian matrix $\boldsymbol{H}_{t,i} = \frac{1}{|\mathcal{D}_i|}\sum_{z_{i,j}\in\mathcal{D}_i} \nabla^2 f\left(\boldsymbol{w}_t, z_{i,j}\right) + \gamma \boldsymbol{I}_d$.
  - Compute local Newton descent direction $\boldsymbol{p}_{t,i} = \boldsymbol{H}_{t,i}^{-1}\boldsymbol{g}_{t,i}$.
  - Encode $\boldsymbol{p}_{t,i}$ as $\boldsymbol{s}_{t,i}$ according to (3.9).
  - Transmit the signal $\boldsymbol{x}_{t,i} = b_{t,i}\boldsymbol{s}_{t,i}$ through wireless channel.
- Edge server receives the signal $\boldsymbol{y}_t$ (3.12) and maintains $\hat{\boldsymbol{p}}_t$ (3.17).
- Edge server performs an update step $\boldsymbol{w}_{t+1} = \boldsymbol{w}_t - \alpha\hat{\boldsymbol{p}}_t$.

3.3.1 Preliminaries

The essence of our proposed algorithm is using local Hessian matrices and local gradients, which is calculated based on the subsets of the total data set, to construct local Newton descent directions and aggregate them. This brings the benefits of reducing the number of communication rounds between the server and the devices. However, since we rely on local information to approximate Newton descent directions, the quality of local Hessian/gradients (i.e., the difference between the local ones and global ones) are of great concern. In order to tackle this issue, we adopt the idea of matrix sketching [7, 26]. Specifically, for a given input matrix $\boldsymbol{M} \in \mathbb{R}^{n\times d}$, we can replace it with $\boldsymbol{C} = \boldsymbol{L}^{\mathsf{T}}\boldsymbol{M} \in \mathbb{R}^{s\times d}$, where matrix $\boldsymbol{C}$ acts as the sketch of $\boldsymbol{M}$ with the sketching matrix $\boldsymbol{L} \in \mathbb{R}^{n\times s}$. In this way, the original problem related to $\boldsymbol{M}$ can be solved more efficiently using the smaller alternative matrix $\boldsymbol{C}$ without losing too much information. The construction of the sketch is similar to the calculation of local Hessian/gradients, where we adopt partial information of the global Hessian/gradients to serve as the local Hessian/gradients. In this chapter, we consider the row sampling scheme in matrix sketching. The sketch $\boldsymbol{C}$ is constructed by the uniform sampled and re-scaled subset of rows of $\boldsymbol{M}$ with sampling probability $\mathbb{P}\left(\boldsymbol{c}_i = \frac{\boldsymbol{m}_j}{\sqrt{sp}}\right) = p,\ p = \frac{1}{n}$, where $\boldsymbol{c}_i$ and $\boldsymbol{m}_j$ are the i-th row of $\boldsymbol{C}$ and j-th row of $\boldsymbol{M}$, respectively. Consequently, the sketching matrix $\boldsymbol{L}$ has only one non-zero entry in each column, and we shall measure the difference between the local Hessian/gradients and global ones with the help of such sketching matrices.

In the following, we consider a linear predictor model $\ell : \mathbb{R} \to \mathbb{R}$, which is frequently used in ML research, e.g., logistic and linear regression, support vector machines, neural networks and graphical models. The function $f\left(\boldsymbol{w}, z_{i,j}\right)$ can thus be rewritten as $\ell\left(\boldsymbol{w}^{\mathsf{T}}\boldsymbol{u}_{i,j}\right)$. Accordingly, we define $\boldsymbol{M}_t = \left[\boldsymbol{m}_1^{\mathsf{T}}, \ldots, \boldsymbol{m}_n^{\mathsf{T}}\right]^{\mathsf{T}} \in \mathbb{R}^{n\times d}$ with $\boldsymbol{m}_{t,j} = \sqrt{\ell''\left(\boldsymbol{w}^{\mathsf{T}}\boldsymbol{u}_{i,j}\right)/n}\boldsymbol{u}_{i,j} \in \mathbb{R}^d$, so the global Hessian matrix can be represented as $\boldsymbol{H}_t = \boldsymbol{M}_t^{\mathsf{T}}\boldsymbol{M}_t + \gamma\boldsymbol{I}_d$. Moreover, by defining $\boldsymbol{N}_t = [\boldsymbol{n}_1, \ldots, \boldsymbol{n}_n] \in$

$\mathbb{R}^{d\times n}$ with $\boldsymbol{n}_i = \nabla f(\boldsymbol{w}_t, z_i)$, the global gradient $\nabla F(\boldsymbol{w}_t)$ can be denoted by $\boldsymbol{g}_t = \frac{1}{n}\boldsymbol{N}_t\boldsymbol{1}+\gamma\boldsymbol{w}_t$. Let $\{\boldsymbol{L}_i\}_{i=1}^m$ be the sketching matrices, the local Hessian matrices and local gradients can be reformulated as:

$$\boldsymbol{H}_{t,i} = \boldsymbol{M}_t^{\mathsf{T}}\boldsymbol{L}_i\boldsymbol{L}_i^{\mathsf{T}}\boldsymbol{M}_t + \gamma\boldsymbol{I}_d\,, \quad \boldsymbol{g}_{t,i} = \frac{1}{n}\boldsymbol{N}_t\boldsymbol{L}_i\boldsymbol{L}_i^{\mathsf{T}}\boldsymbol{1} + \gamma\boldsymbol{w}\,. \tag{3.18}$$

In addition, we define an auxiliary quadratic function as follows to facilitate our analysis:

$$\phi(\boldsymbol{p}) = \frac{1}{2}\boldsymbol{p}^{\mathsf{T}}\boldsymbol{H}_t\boldsymbol{p} - \boldsymbol{g}_t^{\mathsf{T}}\boldsymbol{p} = \frac{1}{2}\boldsymbol{p}^{\mathsf{T}}\left(\boldsymbol{M}_t^{\mathsf{T}}\boldsymbol{M}_t + \gamma\boldsymbol{I}_d\right)\boldsymbol{p} - \boldsymbol{g}_t^{\mathsf{T}}\boldsymbol{p}\,. \tag{3.19}$$

As a quadratic function, the minimum point of $\phi(\boldsymbol{p})$ denoted by $\boldsymbol{p}^*$ can be analytically obtained, which is the same as the exact Newton descent direction vector in (3.3), i.e.,

$$\boldsymbol{p}^* = \arg\min\phi(\boldsymbol{p}) = \nabla^2F^{-1}(\boldsymbol{w}_t)\nabla F(\boldsymbol{w}_t) = \boldsymbol{H}_t^{-1}\boldsymbol{g}_t = \left(\boldsymbol{M}_t^{\mathsf{T}}\boldsymbol{M}_t + \gamma\boldsymbol{I}_d\right)^{-1}\boldsymbol{g}_t\,. \tag{3.20}$$

Due to the effect of channel noise, device selection, and the use of local Newton step, the actual descent direction rather than the exact Newton step $\boldsymbol{p}^*$ is given by:

$$\hat{\boldsymbol{p}}_t = \boldsymbol{p}^* + \underbrace{(\bar{\boldsymbol{p}}_t - \boldsymbol{p}^*)}_{\text{Local Hessian}} + \underbrace{(\boldsymbol{p}_t - \bar{\boldsymbol{p}}_t)}_{\text{Local Gradient}} + \underbrace{(\tilde{\boldsymbol{p}}_t - \boldsymbol{p}_t)}_{\text{Device Selection}} + \underbrace{(\hat{\boldsymbol{p}}_t - \tilde{\boldsymbol{p}}_t)}_{\text{Channel Noise}}\,, \tag{3.21}$$

where $\hat{\boldsymbol{p}}_t$ and $\tilde{\boldsymbol{p}}_t$ are defined in (3.17) and (3.7), respectively, $\boldsymbol{p}_t = \frac{1}{\sum_{i\in\mathcal{S}}|\mathcal{D}_i|}\sum_{i\in\mathcal{S}}|\mathcal{D}_i|\,\boldsymbol{p}_{t,i}$ is the averaged local descent direction vector without device selection, and $\bar{\boldsymbol{p}}_t = \frac{1}{\sum_{i\in\mathcal{S}}|\mathcal{D}_i|}\sum_{i\in\mathcal{S}}|\mathcal{D}_i|\,\bar{\boldsymbol{p}}_{t,i} = \frac{1}{\sum_{i\in\mathcal{S}}|\mathcal{D}_i|}\sum_{i\in\mathcal{S}}|\mathcal{D}_i|\,\boldsymbol{H}_{t,i}^{-1}\boldsymbol{g}_t$ is the Newton descent direction with the exact global gradient. In the following analysis, we will use the quadratic function (3.19) to illustrate how close $\hat{\boldsymbol{p}}_t$ and $\boldsymbol{p}^*$ are. Besides, the error of model parameter vector in the iterates $\boldsymbol{\Delta}_t = \boldsymbol{w}_t - \boldsymbol{w}^*$ acts as the metric, where $\boldsymbol{w}^*$ denotes the optimal solution.

3.3.2 Convergence Analysis

Throughout this chapter, we consider the following assumptions, which are widely adopted in FEEL scenarios [4, 18, 28].

Assumption 1 The global loss function F is L-smooth.

Assumption 2 The global loss function F is strongly convex, which indicates the unique optimal model parameter vector $\boldsymbol{w}^*$ of the FEEL task.

Assumption 3 The local loss function F_i is twice-differentiable, smooth and convex.

Since the local gradients and Hessian matrices are adopted to approximate the global descent direction, the gap between the local direction and the global direction is essential for the convergence analysis. Therefore, we first recall two lemmas to reveal their relationships.

Lemma 3.1 ([11, Variant of Lemma 2]) *Let* $\lambda, \delta = \sum_{i=1}^{m} \delta_i, \{\delta_i\} \in (0, 1)$ *be fixed parameters,* $r = \operatorname{rank}(\boldsymbol{M}_t)$*, and* $\boldsymbol{U} \in \mathbb{R}^{n\times r}$ *be the orthonormal bases of the matrix* $\boldsymbol{M}_t$*. Let* $\mu \in \left[1, \frac{n}{d}\right]$ *be the coherence of* $\boldsymbol{M}_t$ *defined in [30]. Let* $\left\{\boldsymbol{L}_i \in \mathbb{R}^{n\times|\mathcal{D}_i|}\right\}_{i=1}^{m}$ *be independent uniform sampling sketching matrices with* $|\mathcal{D}_i| \geq \frac{3\mu d}{\lambda^2}\log\frac{d}{\delta_i}$*. It holds with the probability exceeding* $1-\delta$ *that:*

$$\left\|\boldsymbol{U}^{\mathsf{T}}\boldsymbol{L}_i\boldsymbol{L}_i^{\mathsf{T}}\boldsymbol{U} - \boldsymbol{I}\right\|_2 \leq \lambda\,, \quad \forall i \in \mathcal{S}\,. \tag{3.22}$$

Lemma 3.2 ([11, Variant of Lemma 3]) *Let* $\left\{\boldsymbol{L}_i \in \mathbb{R}^{n\times|\mathcal{D}_i|}\right\}_{i=1}^{m}$ *be independent uniform sampling sketching matrices,* $\delta = \sum_{i=1}^{m}\delta_i, \{\delta_i\} \in (0, 1)$ *be fixed parameters, then with the probability exceeding* $1-\delta$*, we have:*

$$\left\|\frac{1}{n}\boldsymbol{N}_t\boldsymbol{L}_i\boldsymbol{L}_i^{\mathsf{T}}\boldsymbol{1} - \frac{1}{n}\boldsymbol{N}_t\boldsymbol{1}\right\| \leq \left(1+\sqrt{2\ln\left(\frac{1}{\delta_i}\right)}\right)\sqrt{\frac{1}{|\mathcal{D}_i|}}\max_j\left\|\boldsymbol{n}_j\right\|\,. \tag{3.23}$$

With Lemmas 3.1 and 3.2, we further propose Lemma 3.3 to characterize the gap between $\hat{\boldsymbol{p}}_t$ and $\boldsymbol{p}^*$ via the support quadratic function.

Lemma 3.3 *Let* $\{\boldsymbol{L}_i\}_{i=1}^{m} \in \mathbb{R}^{n\times|\mathcal{D}_i|}$ *be independent uniform sampling sketching matrices,* ϕ_t *be the quadratic function as defined in* (3.19)*,* $\lambda, \{\delta_i\} \in (0, 1)$ *be fixed parameters with* $\tilde{\delta} = \min\{\delta_i\}$ *and* $\hat{\boldsymbol{p}}_t$ *be the approximate descent direction vector defined in* (3.21)*. It holds that:*

$$\phi_t(\boldsymbol{p}^*) \leq \phi_t\left(\hat{\boldsymbol{p}}_t\right) \leq \epsilon^2 + \left(1-\zeta^2\right)\phi_t(\boldsymbol{p}^*)\,,$$

where

$$\zeta^2 = 3\tau^2\left(\lambda + \frac{\lambda^2}{1-\lambda}\right)^2 + 24\vartheta^2\left(\tau\left(\lambda+\frac{\lambda^2}{1-\lambda}\right)+1\right)^2 \tag{3.24}$$

with $\tau = \frac{\sigma_{\max}(\boldsymbol{M}^\top \boldsymbol{M})}{\sigma_{\max}(\boldsymbol{M}^\top \boldsymbol{M})+n\gamma}$, $\vartheta = \max_t \left(1 - \frac{\sum_{i\in\mathcal{S}_t}|\mathcal{D}_i|}{n}\right) < 1$ *and*

$$\begin{aligned}\epsilon^2 = &\frac{3}{\sigma_{\min}(\boldsymbol{H}_t)} \left\| \frac{1}{\left(\sum_{i\in\mathcal{S}_t}|\mathcal{D}_i|\right)\sqrt{\eta_t}} \boldsymbol{a}_t^{\mathsf{H}} \boldsymbol{E}_t \right\|^2 \\ &+ \left[24\left(1 - \frac{\sum_{i\in\mathcal{S}_t}|\mathcal{D}_i|}{n}\right)^2 \frac{1}{\min_{i\in\mathcal{S}_t}|\mathcal{D}_i|} + \frac{m}{n} \right] \\ &\cdot \left[\frac{1}{1-\lambda} \frac{1}{\sqrt{\sigma_{\min}(\boldsymbol{H}_t)}} \left(1 + \sqrt{2\ln\left(\frac{1}{\bar{\delta}}\right)}\right) \max_j \|\boldsymbol{n}_j\| \right]^2 .\end{aligned} \tag{3.25}$$

To illustrate that $\hat{\boldsymbol{p}}_t$ is a good descending direction, we introduce Lemma 3.4 supported by the property of the quadratic function introduced in Lemma 3.3.

Lemma 3.4 ([11, Lemma 6]) *Let* $\zeta \in (0,1)$, ϵ *be any fixed parameter, if* $\hat{\boldsymbol{p}}_t$ *satisfies* $\phi(\hat{\boldsymbol{p}}_t) \le \epsilon^2 + (1-\zeta^2)\min_{\boldsymbol{p}} \phi(\boldsymbol{p})$, *then under Assumption 1, the error of model parameter vector* $\boldsymbol{\Delta}_t = \boldsymbol{w}_t - \boldsymbol{w}^*$ *in iterations satisfies*

$$\boldsymbol{\Delta}_{t+1}^{\mathsf{T}} \boldsymbol{H}_t \boldsymbol{\Delta}_{t+1} \le L \|\boldsymbol{\Delta}_{t+1}\| \|\boldsymbol{\Delta}_t\|^2 + \frac{\zeta^2}{1-\zeta^2} \boldsymbol{\Delta}_t^{\mathsf{T}} \boldsymbol{H}_t \boldsymbol{\Delta}_t + 2\epsilon^2 , \tag{3.26}$$

Based on Lemmas 3.3 and 3.4, we derive the main result as follows.

Theorem 3.1 *Suppose the size of local dataset at each device* $|\mathcal{D}_i| \ge \frac{3\mu d}{\lambda^2}\log\frac{d}{\delta_i}$ *for some* $\lambda, \delta_i \in (0,1)$, *then under Assumption 1 with the probability exceeding* $1-\delta$ *we have*

$$\mathbb{E}(\|\boldsymbol{\Delta}_{t+1}\|) \le \max\left\{ \sqrt{\kappa_t \left(\frac{\zeta^2}{1-\zeta^2}\right)} \|\boldsymbol{\Delta}_t\|, \frac{L}{\sigma_{\min}(\boldsymbol{H}_t)} \|\boldsymbol{\Delta}_t\|^2 \right\} + \epsilon' ,$$

where the expectation takes with respect to the channel noise $\boldsymbol{e}_t$, ζ *is defined as* (3.24), $\kappa_t = \frac{\sigma_{\max}(\boldsymbol{H}_t)}{\sigma_{\min}(\boldsymbol{H}_t)}$ *denotes the condition number of* $\boldsymbol{H}_t$, *and*

$$\begin{aligned}\epsilon' = &\frac{2\sqrt{3}}{\sigma_{\min}(\boldsymbol{H}_t)} \frac{d\sigma}{\sum_{i\in\mathcal{S}_t}|\mathcal{D}_i|} \frac{\|\boldsymbol{a}_t\|}{\sqrt{\eta_t}} + \sqrt{24\left(1 - \frac{\sum_{i\in\mathcal{S}_t}|\mathcal{D}_i|}{n}\right)^2 \frac{1}{\min_{i\in\mathcal{S}_t}|\mathcal{D}_i|} + \frac{m}{n}} \\ &\cdot \frac{1}{1-\lambda} \frac{2}{\sigma_{\min}(\boldsymbol{H}_t)} \left(1 + \sqrt{2\ln\left(\frac{1}{\bar{\delta}}\right)}\right) \max_j \|\boldsymbol{n}_j\| .\end{aligned}$$

According to the analysis, it can be seen that the term $\|\mathbf{\Delta}_t\| = \|\boldsymbol{w}_t - \boldsymbol{w}^*\|$ keeps the property of $\mathbb{E}\left(\|\mathbf{\Delta}_{t+1}\|\right) \leq \max\left\{\omega_1 \|\mathbf{\Delta}_t\|, \omega_2 \|\mathbf{\Delta}_t\|^2\right\} + \epsilon'$. When $\|\mathbf{\Delta}_t\| > \frac{\omega_1}{\omega_2}$, this property can be simplified as $\mathbb{E}\left(\|\mathbf{\Delta}_{t+1}\|\right) \leq \omega_2 \|\mathbf{\Delta}_t\|^2 + \epsilon'$. It is obvious that the proposed algorithm keeps the same quadratic convergence rate as the canonical Newton's method. The algorithm converges to the neighbor of the optimal point quickly. When $\|\mathbf{\Delta}_t\| < \frac{\omega_1}{\omega_2}$, this property turns into $\mathbb{E}\left(\|\mathbf{\Delta}_{t+1}\|\right) \leq \omega_1 \|\mathbf{\Delta}_t\| + \epsilon'$, which means that when $\|\mathbf{\Delta}_t\|$ is small enough, it degenerates into the linear convergence rate. In conclusion, the proposed algorithm keeps a linear-quadratic convergence rate and performs better than the first-order algorithms.

Notice that there is an error term ϵ' in each iteration, which comes from the approximation, device selection and channel noise. Consider the noise-free case without device selection (i.e., $\sigma = 0$ and $|\mathcal{S}_t| = m$), the error term degenerates to

$$\epsilon_1' = \frac{1}{1-\lambda} \frac{2}{\sigma_{\min}\left(\boldsymbol{H}_t\right)} \left(1 + \sqrt{2 \ln\left(\frac{1}{\bar{\delta}}\right)}\right) \sqrt{\frac{m}{n}} \max_j \|\boldsymbol{n}_j\|, \tag{3.27}$$

which is exactly the same as the error term introduced in [11]. With the algorithm executed iteratively, the gap between the expected global loss function value and the optimal one is upper bounded by this accumulative error term. Therefore, the active device set $\mathcal{S}_t$, the receive beamforming vectors $\{\boldsymbol{a}_t\}$ and the scaling factors $\{\eta_t\}$ need to be tuned in each iteration so as to reduce the error gap.

3.4 System Optimization

In this section, we first formulate a system optimization problem to minimize the error term in the convergence analysis results. Then, we propose our approach for joint optimization of device selection and receive beamforming vector.

3.4.1 *Problem Formulation*

In light of convergence analysis, to obtain a precise model parameter vector, minimizing the error gap demonstrated in Theorem 3.1 is a key issue. It is observed that the coefficients of the two iterative terms $\mathbf{\Delta}_t$ and $\mathbf{\Delta}_{t+1}$ in Theorem 3.1 are independent of variables $\mathcal{S}_t$, $\boldsymbol{a}_t$ and η_t. Therefore, in order to achieve the minimization of the total error gap, we only need to minimize the error term ϵ' in each iteration as follows

$$\begin{aligned}
\min_{\mathcal{S}_t,\boldsymbol{a}_t,\eta_t} \quad & \frac{2\sqrt{3}}{\sigma_{\min}\left(\boldsymbol{H}_t\right)} \frac{d\sigma}{\sum_{i\in\mathcal{S}_t}|\mathcal{D}_i|} \frac{\|\boldsymbol{a}_t\|}{\sqrt{\eta_t}} \\
& + \sqrt{24\left(1-\frac{\sum_{i\in\mathcal{S}_t}|\mathcal{D}_i|}{n}\right)^2 \frac{1}{\min_{i\in\mathcal{S}_t}|\mathcal{D}_i|}+\frac{m}{n}} \\
& \cdot \frac{1}{1-\lambda}\frac{2}{\sigma_{\min}\left(\boldsymbol{H}_t\right)}\left(1+\sqrt{2\ln\left(\frac{1}{\tilde{\delta}}\right)}\right)\max_j\|\boldsymbol{n}_j\| \\
\text{s.t.} \quad & \frac{\eta_t}{\left\|\boldsymbol{a}_t^{\mathsf{H}}\tilde{\boldsymbol{h}}_{t,i}\right\|^2} \leq dP_0 \quad \forall i\in\mathcal{S}_t
\end{aligned} \tag{3.28}$$

The power constraint in (3.28) can be rewritten in the form of the restriction of scaling factor: $\eta_t \leq dP_0\left\|\boldsymbol{a}_t^{\mathsf{H}}\tilde{\boldsymbol{h}}_{t,i}\right\|^2$, $\forall i\in\mathcal{S}_t$. We take the negative correlation between the scaling factor η_t and the objective function value ϵ' into consideration. η_t can be set as $\eta_t = dP_0\min_{i\in\mathcal{S}_t}\left\|\boldsymbol{a}_t^{\mathsf{H}}\tilde{\boldsymbol{h}}_{t,i}\right\|^2$ [3], and the problem can be simplified as $\mathscr{P}$:

$$\begin{aligned}
\mathscr{P}: \min_{\mathcal{S}_t,\boldsymbol{a}_t} \quad & \frac{\sqrt{3}d\sigma}{\sqrt{P_0}\sum_{i\in\mathcal{S}_t}|\mathcal{D}_i|} \max_{i\in\mathcal{S}_t}\left(\frac{\|\boldsymbol{a}_t\|}{\left\|\boldsymbol{a}_t^{\mathsf{H}}\tilde{\boldsymbol{h}}_{t,i}\right\|}\right) \\
& + \sqrt{24\left(1-\frac{\sum_{i\in\mathcal{S}_t}|\mathcal{D}_i|}{n}\right)^2 \frac{1}{\min_{i\in\mathcal{S}_t}|\mathcal{D}_i|}+\frac{m}{n}} \\
& \cdot \frac{1}{1-\lambda}\frac{2}{\sigma_{\min}\left(\boldsymbol{H}_t\right)}\left(1+\sqrt{2\ln\left(\frac{1}{\tilde{\delta}}\right)}\right)\max_j\|\boldsymbol{n}_j\| .
\end{aligned} \tag{3.29}$$

We have the following key observations for solving (3.29):

- Intuitively, to achieve the minimization of the objective value of $\mathscr{P}$, the number of selected devices is supposed to be maximized, then $\mathscr{P}$ will degenerate into the form of traditional beamforming optimization. However, the term $\max_{i\in\mathcal{S}_t}\left(\frac{\|\boldsymbol{a}_t\|}{\left\|\boldsymbol{a}_t^{\mathsf{H}}\tilde{\boldsymbol{h}}_{t,i}\right\|}\right)$ is related to device selection, which further results in the incorrectness to directly maximize $|\mathcal{S}_t|$.
- By searching over all the possible participating device sets, the optimal $\mathcal{S}_t$ can be determined. Still, the number of devices m can be very large, leading to an exponential growth of the optimization procedure in the number of devices m.

- After the search of participating devices, the remaining problem is a typical beamforming optimization problem, but it is still non-convex and intractable.

3.4.2 Receive Beamforming Optimization

For a given set of selected devices $\mathcal{S}_t$, $\mathscr{P}$ can be simplified as

$$\mathscr{P}_1 : \min_{\boldsymbol{a}_t} \max_{i \in \mathcal{S}_t} \frac{\|\boldsymbol{a}_t\|}{\left|\boldsymbol{a}_t^{\mathsf{H}} \tilde{\boldsymbol{h}}_{t,i}\right|},$$

which is equivalent to: $\min_{\boldsymbol{a}_t} \max_{i \in \mathcal{S}_t} \frac{\|\boldsymbol{a}_t\|^2}{\left|\boldsymbol{a}_t^{\mathsf{H}} \tilde{\boldsymbol{h}}_{t,i}\right|^2}$. This can be further reformulated as $\mathscr{P}_1'$ according to the analysis in [3]:

$$\mathscr{P}_1' : \min_{\boldsymbol{a}_t} \quad \|\boldsymbol{a}_t\|^2 \qquad \text{s.t.} \quad \left|\boldsymbol{a}_t^{\mathsf{H}} \tilde{\boldsymbol{h}}_{t,i}\right|^2 \geq 1 \quad \forall i \in \mathcal{S}_t \ .$$

It can be seen that $\mathscr{P}_1'$ is actually a quadratically constrained quadratic programming problem, which is difficult to solve. We first use the matrix lifting technique to pre-process $\mathscr{P}_1'$ and turn it into a low-rank optimization form. Specifically, let $\boldsymbol{A} = \boldsymbol{a}_t \boldsymbol{a}_t^{\mathsf{H}}$ with $\text{rank}(\boldsymbol{A}) = 1$ and $\boldsymbol{Q}_i = \tilde{\boldsymbol{h}}_{t,i} \tilde{\boldsymbol{h}}_{t,i}^{\mathsf{H}}$, $\mathscr{P}_1'$ can be recast as:

$$\min_{\boldsymbol{A}} \quad \text{Tr}(\boldsymbol{A}) \qquad \text{s.t.} \quad \boldsymbol{A} \succeq \boldsymbol{0}, \ \ \text{rank}(\boldsymbol{A}) = 1, \ \ \text{Tr}\left(\boldsymbol{A}\boldsymbol{Q}_i\right) \geq 1 \quad \forall i \in \mathcal{S}_t \ .$$

To guarantee the rank-one constraint, we can replace it with its equivalent form [13, 29]: $\text{Tr}(\boldsymbol{A}) - \|\boldsymbol{A}\|_2 = 0$ with $\text{Tr}(\boldsymbol{A}) > 0$. Then, the original problem turns into a difference-of-convex-function (DC) program. By solving this DC program, a more precise solution can be obtained since all constraints are satisfied. Therefore, we develop a DC Algorithm (DCA) based on the principles in [14, 24] to solve this problem. Specifically, we can get the following problem by taking the new constraint as a penalty term:

$$\begin{aligned} &\min_{\boldsymbol{A}} \quad \text{Tr}(\boldsymbol{A}) + \theta\left(\text{Tr}(\boldsymbol{A}) - \|\boldsymbol{A}\|_2\right) \quad \text{s.t.} \quad \boldsymbol{A} \succeq \boldsymbol{0}, \\ &\text{Tr}(\boldsymbol{A}) > 0, \ \text{Tr}\left(\boldsymbol{A}\boldsymbol{Q}_i\right) \geq 1 \ \ \forall i \in \mathcal{S}_t \ , \end{aligned}$$

where θ is the penalty factor. Although this is still a non-convex problem owing to the concave term $-\|\boldsymbol{A}\|_2$, we can take the linearization of $\|\boldsymbol{A}\|_2$ and convert it into a convex subproblem:

$$\mathscr{P}_{DCA} : \min_{\boldsymbol{A}} (1+\theta)\,\text{Tr}(\boldsymbol{A}) - \theta\left\langle \partial \left\|\boldsymbol{A}_j\right\|_2, \boldsymbol{A}\right\rangle \tag{3.30}$$

$$\text{s.t. } \boldsymbol{A} \succeq \boldsymbol{0},\ \operatorname{Tr}(\boldsymbol{A}) > 0,\ \operatorname{Tr}\left(\boldsymbol{A}\boldsymbol{Q}_i\right) \geq 1\ \forall i \in \mathcal{S}_t, \tag{3.31}$$

where $\langle \cdot, \cdot \rangle$ denotes the inner product of two matrices and $\partial \left\| \boldsymbol{A}_j \right\|_2$ represents the subgradient of $\left\| \boldsymbol{A}_j \right\|_2$ at $\boldsymbol{A}_j$. Therefore, the result can be obtained by iteratively solving $\mathscr{P}_{DCA}$ until $\operatorname{Tr}(\boldsymbol{A}) - \|\boldsymbol{A}\|_2$ is sufficiently small. The overall procedure of DCA is as summarized in Algorithm 4.

Algorithm 4: DC algorithm for receiver beamforming optimization (DCA)

input: Effective channel coefficients $\left\{\tilde{\boldsymbol{h}}_{t,i}\right\}$, penalty factor θ, threshold ξ
Turn $\mathscr{P}$ into the DCA form $\mathscr{P}_{DCA}$.
Choose $\boldsymbol{A}_0 \succeq 0$, set $j = 1$.
while $\left|\operatorname{Tr}\left(\boldsymbol{A}_{j-1}\right) - \left\|\boldsymbol{A}_{j-1}\right\|_2\right| \geq \xi$ **do**
  Compute the subgradient $\partial \left\|\boldsymbol{A}_{j-1}\right\|_2$.
  Substitute $\partial \left\|\boldsymbol{A}_{j-1}\right\|_2$ into $\mathscr{P}_{DCA}$, solve the subproblem and set the result as $\boldsymbol{A}_j$.
  $j \leftarrow j + 1$.

Algorithm 5: System optimization approach GS+DCA

input: Effective channel coefficients $\left\{\tilde{\boldsymbol{h}}_{t,i}\right\}$, $T^{(0)}$, ρ, K
output: $\mathcal{S}^{(k+1)}$ and its corresponding $\boldsymbol{a}^{(K+1)}$.
Initialization: $\mathcal{S}^{(0)} = \mathcal{S}$
for *iteration* $k = 0, 1, 2, ..., K$ **do**
  Generate the neighboring solution set $\mathcal{F}^{(k)}$.
  for *each* $\tilde{\mathcal{S}} \in \mathcal{F}^{(k)}$ **do**
    Substitute $\tilde{\mathcal{S}}$ into $\mathscr{P}_1$, then solve the problem using DCA to get the corresponding optimal $\tilde{\boldsymbol{a}}$.
  Sample $\tilde{\mathcal{S}}^{(k)}$ according to the probability $\mathbb{P}\left(\tilde{\mathcal{S}}^{(k)}\right) = \frac{\exp\left(-J\left(\tilde{\mathcal{S}}^{(k)},\tilde{\boldsymbol{a}}^{(k)}\right)/T^{(k)}\right)}{\sum_{\tilde{\mathcal{S}}\in\mathcal{F}^{(k)}} \exp\left(-J\left(\tilde{\mathcal{S}},\tilde{\boldsymbol{a}}\right)/T^{(k)}\right)}$.
  $\mathcal{S}^{(k+1)} \leftarrow \tilde{\mathcal{S}}^{(k)}$, $T^{(k+1)} \leftarrow \rho T^{(k)}$.

3.4.3 Device Selection Optimization

As mentioned above, the device selection is a combinatorial optimization problem, which is impossible to perform a traversal in the whole solution space. Thus, we adopt the well-known Gibbs Sampling (GS) [10] method to optimize the selection of device set iteratively. The main idea of GS is that in each iteration, a device set is sampled from the neighbors of the current device set according to an appropriate distribution. In this way, the set of selected devices can gradually approach the global optimal solution.

To be specific, we treat different sets of selected devices as states, and the goal is to find the state which can minimize the objective value in $\mathscr{P}$. For the sake of such state, at iteration k of GS's process, with the set of selected devices $\mathcal{S}^{(k-1)}$ given in the last iteration, we first generate the neighboring solution set of $\mathcal{S}^{(k-1)}$. The neighboring solution set, denoted by $\mathcal{F}^{(k)}$, contains the device sets that differ from the $\mathcal{S}^{(k-1)}$ in only one entry. For example, by assuming $\mathcal{S} = \{0, 1, 2\}$ and $\mathcal{S}^{(k-1)} = \{1, 2\}$, then we have $\mathcal{F}^{(k)} = \{\{0, 1, 2\}, \{2\}, \{1\}\}$.

After the identification of the neighboring solution set, the candidate states are also determined according to the sets in $\mathcal{F}^{(k)}$, and we need to choose a state to approach the optimal set. Based on the distribution introduced in [2], we sample a device set in $\mathcal{F}^{(k)}$ with the probability

$$\mathbb{P}\left(\tilde{\mathcal{S}}^{(k)}\right) = \frac{\exp\left(-J\left(\tilde{\mathcal{S}}^{(k)}, \tilde{\boldsymbol{a}}^{(k)}\right)/T^{(k)}\right)}{\sum_{\tilde{\mathcal{S}}\in\mathcal{F}^{(k)}} \exp\left(-J\left(\tilde{\mathcal{S}}, \tilde{\boldsymbol{a}}\right)/T^{(k)}\right)}, \tag{3.32}$$

where $J(\boldsymbol{x}, \boldsymbol{y})$ denotes the objective function value of $\mathscr{P}$. Here, the receiver beamforming vector is calculated through DCA with the given set of selected devices.

In (3.32), there is a special parameter $T^{(k)}$ serving as the temperature. The algorithm starts from a relatively high temperature $T^{(0)}$ in order to move around the solution space freely, rather than being stuck in a local minimum point. As the algorithm proceeds, the algorithm slowly decreases the temperature by the factor ρ to focus on the states that minimize the objective function. Besides, to reduce the computational complexity, we have adopted a similar warm start technique as in [18]. The optimal beamforming vector in the previous iteration is used to serve as the initial point to accelerate the process of beamforming optimization. The overall process of system optimization is outlined in Algorithm 5.

3.5 Simulation Results

3.5.1 Simulation Settings

In this section, we evaluate the performance of the proposed schemes to demonstrate the advantage of our proposed second-order federated optimization algorithm and the effectiveness of our system optimization approach. Code for our experiments are available at: https://github.com/Golden-Slumber/AirFL-2nd. We first consider logistic regression with the loss function of the i-th device

$$F_i(\boldsymbol{w}) = \frac{1}{|\mathcal{D}_i|} \sum_{z_{i,j}=(\boldsymbol{u}_{i,j}, v_{i,j})\in\mathcal{D}_i} \log\left(1 + \exp\left(-v_{i,j}\boldsymbol{u}_{i,j}^{\mathsf{T}}\boldsymbol{w}\right)\right) + \frac{\gamma}{2}\|\boldsymbol{w}\|_2^2,$$

where the regularization parameter is set to be $\gamma = 10^{-8}$. As for datasets, we adopt four different standard datasets from the LIBSVM library: Covtype, a9a, w8a, and phishing. In this chapter, we consider a distributed wireless scenario, where these data samples are uniformly distributed in $m = 20$ devices, the server is equipped with $k = 5$ antennas. The channel coefficients are given by the small-scale fading coefficients $\{\boldsymbol{h}'_{t,i}\}$ multiplied by the path loss gain PL_i, i.e., $\boldsymbol{h}_{t,i} = PL_i \boldsymbol{h}'_{t,i}$. Here, the small-scale fading coefficients follow the i.i.d complex normal distribution $\mathcal{CN}(0, \boldsymbol{I})$. The path loss gain is given by $PL_i = \sqrt{G_0}\,(d_0/d_i)^{\nu/2}$, where $G_0 = 10^{-3.35}$ is the average channel power gain with the distance to the server $d_0 = 1$ m, $d_i \in [100, 120]$ stands for the distance between the i-th device and the server, and $\nu = 3.76$ represents the path loss exponent factor. For the step size α, we use backtracking line search to find α satisfying the Armijo–Goldstein condition [21, Chapter 3]. For the system optimization, we set $\lambda = 0.1$, $\tilde{\delta} = 0.01$, penalty factor $\theta = 1$, threshold $\xi = 10^{-10}$, initial temperature $T_0 = 100$, $\rho = 0.9$, and $K = 30$. Besides, we use Baseline 0 to denote the centralized training setting in all experiments.

Furthermore, we also consider an image classification problem on a non-i.i.d dataset constructed from the Fashion-MNIST dataset at the end of this section. To address it, we train a softmax classifier with cross-entropy loss and ℓ_2 regularization term. To be specific, the loss function of the i-th device is given as $F_i(\boldsymbol{W}) = \frac{1}{|\mathcal{D}_i|}\sum_{(\boldsymbol{u},v)\in\mathcal{D}_i}\sum_{c=1}^{C} \mathbf{1}\{v = c\}\log\frac{\exp(\boldsymbol{u}^{\mathsf{T}}\boldsymbol{w}_c)}{\sum_{j=1}^{C}\exp(\boldsymbol{u}^{\mathsf{T}}\boldsymbol{w}_j)} + \frac{\gamma}{2}\sum_{c=1}^{C}\|\boldsymbol{w}_c\|_2^2$, where $\boldsymbol{W} = [\boldsymbol{w}_1, \ldots, \boldsymbol{w}_C]$ is the concatenation of parameter vectors related to different classes, and $C = 10$ represents the total number of classes.

3.5.2 *Effectiveness of Proposed System Optimization Approach*

In this experiment, we evaluate the performance using GS+DCA to accomplish system optimization with four settings:

1. Perfect aggregation, where the model is aggregated without wireless channel impact.
2. GS+SDR, where the receiver beamforming optimization is performed through SDR.
3. DCA only, where we only perform beamforming optimization through DCA.
4. SDR only, where we only perform beamforming optimization through SDR.

To verify the effectiveness of the device selection, we consider the distance heterogeneity and data size heterogeneity in this experiment. Specifically, as for distance heterogeneity, we set the distance of 10% devices to be $d_i \in [200, 220]$ while the rest to be $d_i \in [50, 60]$. As for data size heterogeneity, we set the data size of 10% devices to be $|\mathcal{D}_i| \in [0.008\frac{n}{m}, 0.01\frac{n}{m}]$ while the rest to be $|\mathcal{D}_i| \in [1.01\frac{n}{m}, 1.11\frac{n}{m}]$.

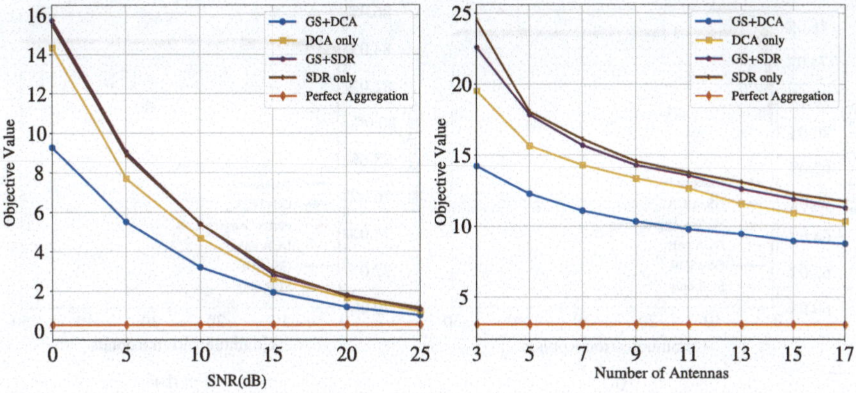

Fig. 3.2 Objective value of system optimization problem $\mathscr{P}$ versus SNR and number of antennas

We first numerically evaluate the objective value of the system optimization problem $\mathscr{P}$ under different settings in Fig. 3.2 by averaging 100 channel realizations. The objective value of perfect aggregation does not depend on SNR and the number of antennas since the error during the FEEL process in this situation only comes from the approximation as (3.27) indicates. The objective values of all settings decrease as SNR and the number of antennas increase, due to the mitigation of noise effect and the increase of diversity gain [3], respectively. However, the objective value of GS+DCA is smaller than that of other settings. On the one hand, SDR fails to give a precise solution for the receiver beamforming vector as the size of the problem grows. This further leads to the ineffectiveness of device selection in GS+SDR and worse performance compared with the settings using DCA to perform beamforming optimization. On the other hand, device selection in GS+DCA mitigates the straggler issue caused by distance heterogeneity and data size heterogeneity, resulting in a better performance compared with DCA only.

Figure 3.3 plots the training loss and test accuracy for our proposed algorithm in different system optimization settings, where SNR is set to 35 dB. The results show that with device selection and a more precise solution given by DCA, the error term can be minimized in each iteration and a smaller optimality gap close to that of perfect aggregation can be obtained. This smaller optimality gap further leads to higher test accuracy, demonstrating that our proposed system optimization approach effectively improves learning performance.

3.6 Summary

In this chapter, we studied the communication issues of federated learning in a wireless environment. In order to improve the communication efficiency and the learning performance of wireless FEEL, we proposed a communication-efficient

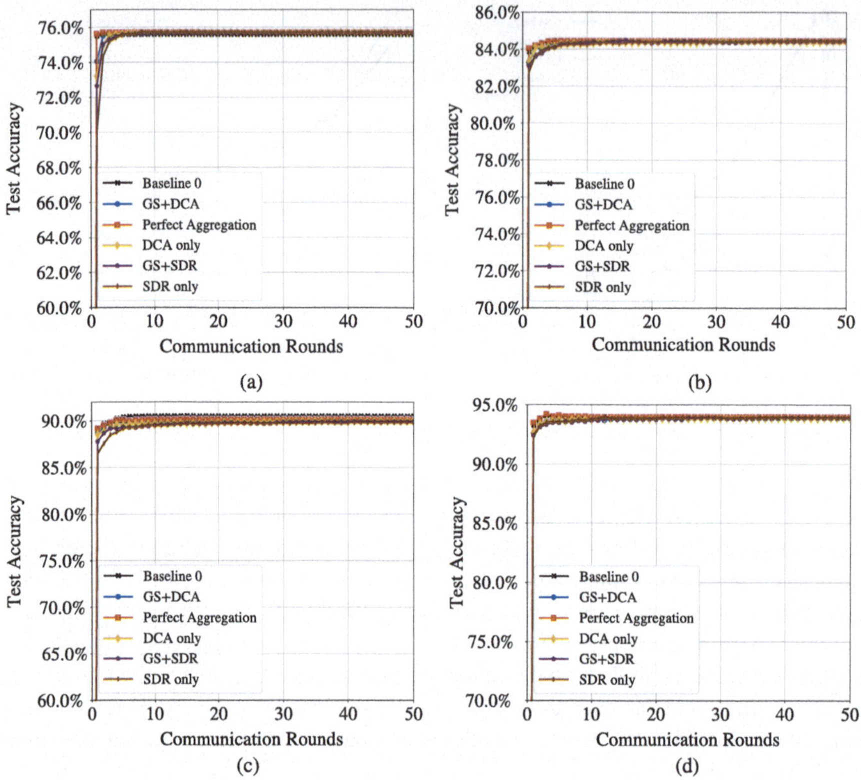

Fig. 3.3 Test accuracy of the proposed algorithm in different system optimization settings. (**a**) Covtype. (**b**) a9a. (**c**) w8a. (**d**) Phishing

federated learning algorithm. Specifically, we adopted the second-order federated optimization approach together with the over-the-air computation technique. Furthermore, we provided a theoretical analysis of the convergence behavior of the proposed algorithm. The results revealed that it achieves a linear-quadratic convergence rate, which outperforms first-order algorithms. Besides, as the proposed algorithm is accompanied by an accumulative error term in each iteration, a system optimization problem was formulated aiming to minimize the total error gap and achieve a precise model. We then developed an efficient approach to jointly optimize device selection and receiving beamformer. The experimental results prove that our proposed algorithm and system optimization approach achieve a better performance than the compared methods.

References

1. S. Bischoff, S. Günnemann, M. Jaggi, and S. U. Stich. On second-order optimization methods for federated learning. *arXiv preprint arXiv:2109.02388*, 2021.
2. P. Brémaud. *Markov chains: Gibbs fields, Monte Carlo simulation, and queues*, volume 31. Springer Science & Business Media, 2013.
3. L. Chen, X. Qin, and G. Wei. A uniform-forcing transceiver design for over-the-air function computation. *IEEE Wireless Commun. Lett.*, 7(6):942–945, 2018.
4. M. Chen, Z. Yang, W. Saad, C. Yin, H. V. Poor, and S. Cui. A joint learning and communications framework for federated learning over wireless networks. *IEEE Trans. Wirel. Commun.*, 20(1):269–283, 2020.
5. R. Crane and F. Roosta. Dingo: Distributed newton-type method for gradient-norm optimization. *arXiv preprint arXiv:1901.05134*, 2019.
6. R. Crane and F. Roosta. Dino: Distributed newton-type optimization method. In *ICML*, pages 2174–2184. PMLR, 2020.
7. P. Drineas and M. W. Mahoney. Randnla: randomized numerical linear algebra. *Communications of the ACM*, 59(6):80–90, 2016.
8. W. Fang, M. Fu, K. Wang, Y. Shi, and Y. Zhou. Stochastic beamforming for reconfigurable intelligent surface aided over-the-air computation. In *GLOBECOM 2020–2020 IEEE Global Communications Conference*, pages 1–6. IEEE, 2020.
9. W. Fang, Y. Jiang, Y. Shi, Y. Zhou, W. Chen, and K. B. Letaief. Over-the-air computation via reconfigurable intelligent surface. *Trans. Commun.*, 69(12):8612–8626, Dec. 2021.
10. S. Geman and D. Geman. Stochastic relaxation, gibbs distributions, and the bayesian restoration of images. *IEEE PAMI*, (6):721–741, 1984.
11. A. Ghosh, R. K. Maity, and A. Mazumdar. Distributed newton can communicate less and resist byzantine workers. *arXiv preprint arXiv:2006.08737*, 2020.
12. H. Guo, A. Liu, and V. K. Lau. Analog gradient aggregation for federated learning over wireless networks: Customized design and convergence analysis. *IEEE Internet of Things Journal*, 8(1):197–210, 2020.
13. S. Hua, K. Yang, and Y. Shi. On-device federated learning via second-order optimization with over-the-air computation. In *2019 IEEE 90th Vehicular Technology Conference (VTC2019-Fall)*, pages 1–5. IEEE, 2019.
14. K. Khamaru and M. Wainwright. Convergence guarantees for a class of non-convex and non-smooth optimization problems. In *International Conference on Machine Learning*, pages 2601–2610. PMLR, 2018.
15. K. B. Letaief, W. Chen, Y. Shi, J. Zhang, and Y.-J. A. Zhang. The roadmap to 6g: Ai empowered wireless networks. *IEEE Commun. Mag.*, 57(8):84–90, 2019.
16. L. Li, L. Yang, X. Guo, Y. Shi, H. Wang, W. Chen, and K. B. Letaief. Delay analysis of wireless federated learning based on saddle point approximation and large deviation theory. *arXiv preprint arXiv:2103.16994*, 2021.
17. T. Lin, S. U. Stich, K. K. Patel, and M. Jaggi. Don't use large mini-batches, use local sgd. *arXiv preprint arXiv:1808.07217*, 2018.
18. H. Liu, X. Yuan, and Y.-J. A. Zhang. Reconfigurable intelligent surface enabled federated learning: A unified communication-learning design approach. *arXiv preprint arXiv:2011.10282*, 2020.
19. B. McMahan, E. Moore, D. Ramage, S. Hampson, and B. A. y Arcas. Communication-efficient learning of deep networks from decentralized data. In *Artificial Intelligence and Statistics*, pages 1273–1282. PMLR, 2017.
20. B. Nazer and M. Gastpar. Computation over multiple-access channels. *IEEE Trans. Inf. Theory*, 53(10):3498–3516, 2007.
21. J. Nocedal and S. Wright. *Numerical optimization*. Springer Science & Business Media, 2006.
22. R. Pathak and M. J. Wainwright. Fedsplit: An algorithmic framework for fast federated optimization. *arXiv preprint arXiv:2005.05238*, 2020.

23. O. Shamir, N. Srebro, and T. Zhang. Communication-efficient distributed optimization using an approximate newton-type method. In *ICML*, pages 1000–1008. PMLR, 2014.
24. P. D. Tao and L. T. H. An. Convex analysis approach to dc programming: theory, algorithms and applications. *Acta mathematica vietnamica*, 22(1):289–355, 1997.
25. S. Wang, F. Roosta, P. Xu, and M. W. Mahoney. Giant: Globally improved approximate newton method for distributed optimization. In *Adv. Neural Inf. Process. Syst.*, pages 2332–2342, 2018.
26. D. P. Woodruff. Sketching as a tool for numerical linear algebra. *arXiv preprint arXiv:1411.4357*, 2014.
27. B. Woodworth, K. K. Patel, S. Stich, Z. Dai, B. Bullins, B. Mcmahan, O. Shamir, and N. Srebro. Is local sgd better than minibatch sgd? In *ICML*, pages 10334–10343. PMLR, 2020.
28. C. Xiaowen, Z. Guangxu, X. Jie, W. Zhiqin, and C. Shuguang. Optimized power control design for over-the-air federated edge learning. *arXiv preprint arXiv:2106.09316*, 2021.
29. K. Yang, T. Jiang, Y. Shi, and Z. Ding. Federated learning via over-the-air computation. *IEEE Trans. Wirel. Commun.*, 19(3):2022–2035, 2020.
30. K. Yang, Y. Shi, Y. Zhou, Z. Yang, L. Fu, and W. Chen. Federated machine learning for intelligent iot via reconfigurable intelligent surface. *IEEE Network*, 34(5):16–22, 2020.
31. P. Yang, Y. Jiang, T. Wang, Y. Zhou, Y. Shi, and C. N. Jones. Over-the-air federated learning via second-order optimization. *IEEE Trans. Wirel. Commun.*, 21(12):10560–10575, 2022.
32. H. Yuan and T. Ma. Federated accelerated stochastic gradient descent. *arXiv preprint arXiv:2006.08950*, 2020.
33. Y. Zhang and X. Lin. Disco: Distributed optimization for self-concordant empirical loss. In *ICML*, pages 362–370. PMLR, 2015.

Chapter 4
Zeroth-Order Algorithm for Federated Edge Learning

4.1 Overview

Due to the widespread adoption of FEEL, the federated optimization for model training has garnered increasing attention in recent years, capturing the interest of both academia and industry alike. To achieve rapid convergence rates and alleviate communication burdens, researchers have introduced a plethora of algorithms. These algorithms encompass both first-order approaches, such as FedAvg [17], FedPD [25], and FedNova [22], as well as second-order techniques, exemplified by FedDANE [13].

The majority of existing algorithms rely on gradient and/or Hessian information to address federated optimization challenges. Nevertheless, in scenarios where the analytical expressions of loss functions are either unavailable or computationally expensive, obtaining gradients becomes an insurmountable hurdle. These scenarios include applications like federated hyperparameter tuning [3], distributed black-box attacks on deep neural networks [10], and online sensor selection where gradient computations entail complex matrix inversions [16]. In essence, existing algorithms are ill-equipped to efficiently handle federated optimization problems when gradient information is absent.

Furthermore, current distributed zeroth-order algorithms [20, 24] fail to account for the distinctive characteristics of FEEL, where all edge devices actively participate in optimization, executing one local update per communication round. This approach is infeasible in FEEL systems with a substantial number of edge devices and high-dimensional model parameters due to the associated communication overhead.

To address the aforementioned challenges, this chapter proposes a novel FedZO algorithm [5]. The convergence guarantee for the FedZO algorithm and the maximum number of local iterates required for preserving convergence are also investigated. To be implemented over the FEEL system with a large number of edge devices, the FedZO algorithm enables partial device participation for the reduction

Y. Zhou et al., *Federated Edge Learning*, Wireless Networks,
https://doi.org/10.1007/978-3-031-96649-1_4

of communication overhead. Extensive simulations are conducted to evaluate the efficiency and convergence performance of the proposed FedZO algorithm.

4.2 Federated Zeroth-Order Optimization

4.2.1 Problem Formulation

We consider an FEEL task performing in a network comprised of a central server and N edge devices. The central server aims to efficiently coordinate all edge devices for collaboratively addressing the given federated optimization problem

$$\min_{\boldsymbol{x}\in\mathbb{R}^d} f(\boldsymbol{x}) \triangleq \frac{1}{N}\sum_{i=1}^{N} f_i(\boldsymbol{x}), \tag{4.1}$$

where $\boldsymbol{x} \in \mathbb{R}^d$ represents the model parameter with dimension d, and $f_i(\boldsymbol{x})$ and $f(\boldsymbol{x})$ represent the local loss function of edge device i and the global loss function at the central server evaluated at model parameter $\boldsymbol{x}$, respectively. We operate under the assumption that every edge device, each possessing a local dataset, holds equal significance for the overall global model [21]. In (4.1), $f_i(\boldsymbol{x})$ quantifies the expected risk concerning the local data distribution at edge device i, represented as $\mathcal{D}_i$. This is given by

$$f_i(\boldsymbol{x}) \triangleq \mathbb{E}_{\xi_i\sim\mathcal{D}_i}[F_i(\boldsymbol{x},\xi_i)],$$

where $F_i(\boldsymbol{x},\xi_i)$ denotes the loss corresponding to ξ_i evaluated at the model parameter $\boldsymbol{x}$, and ξ_i represents a random variable that is uniformly distributed across $\mathcal{D}_i$. Specifically, each ξ_i is realized by a single data sample. Given the sampling information of ξ_i and $\boldsymbol{x}$, edge device i can obtain the function value of F_i. This serves as a stochastic approximation of the expected loss $f_i(\boldsymbol{x})$. Notably, both the analytic expression and gradient information of F_i are not accessible here.

4.2.2 Preliminaries on Stochastic Gradient Estimator

We utilize a mini-batch-type stochastic gradient estimator for function F_i [15], is defined as

$$\begin{aligned}&\widetilde{\nabla} F_i\left(\boldsymbol{x},\{\xi_{i,m}\}_{m=1}^{b_1},\{\boldsymbol{v}_{i,n}\}_{n=1}^{b_2},\mu\right)\\ &=\frac{1}{b_1b_2}\sum_{m=1}^{b_1}\sum_{n=1}^{b_2}\frac{d\boldsymbol{v}_{i,n}}{\mu}\Big(F_i(\boldsymbol{x}+\mu\boldsymbol{v}_{i,n},\xi_{i,m})-F_i(\boldsymbol{x},\xi_{i,m})\Big),\end{aligned} \tag{4.2}$$

where $\{\xi_{i,m}\}_{m=1}^{b_1}$ is a sequence of independent and identically distributed (i.i.d.) random variables with the same distribution as ξ_i, $\{\boldsymbol{v}_{i,n}\}_{n=1}^{b_2}$ is a sequence of i.i.d. random vectors with uniform distribution over $\mathbb{S}^d$, i.e., $\mathcal{U}(\mathbb{S}^d)$, and μ is a positive step size. It has been shown in [8] that

$$\mathbb{E}\left[\frac{d\boldsymbol{v}_{i,n}}{\mu}\Big(F_i(\boldsymbol{x}+\mu\boldsymbol{v}_{i,n},\xi_{i,m})-F_i(\boldsymbol{x},\xi_{i,m})\Big)\right]=\nabla f_i^{\mu}(\boldsymbol{x}), \tag{4.3}$$

where the expectation is taken over $\{\xi_{i,m}, \boldsymbol{v}_{i,n}\}$ and $f_i^{\mu}(\boldsymbol{x}) = \mathbb{E}_{\boldsymbol{u}\sim\mathcal{U}(\mathbb{B}^d)}[f_i(\boldsymbol{x}+\mu\boldsymbol{u})]$ is a locally averaged version of $f_i(\boldsymbol{x})$. Furthermore, we have

$$\mathbb{E}\left[\widetilde{\nabla} F_i\left(\boldsymbol{x}, \{\xi_{i,m}\}_{m=1}^{b_1}, \{\boldsymbol{v}_{i,n}\}_{n=1}^{b_2}, \mu\right)\right] = \nabla f_i^{\mu}(\boldsymbol{x}). \tag{4.4}$$

where the expectation is taken over $\{\xi_{i,m}\}_{m=1}^{b_1}$ and $\{\boldsymbol{v}_{i,n}\}_{n=1}^{b_2}$. As $\nabla f_i^{\mu}(\boldsymbol{x})$ is a biased approximation of $\nabla f_i(\boldsymbol{x})$ [16], (4.2) is a biased estimate of the actual gradient.

4.2.3 FedZO Algorithm

Inspired by the FedAvg algorithm, we have devised a FedZO optimization algorithm, outlined in Algorithm 6. The core objective of the FedZO algorithm is to remove reliance on the gradient and reduce the frequency of model exchanges. To this end, we employ a gradient estimator as in Eq. (4.2) and execute H steps of stochastic zeroth-order updates per communication round. The FedZO algorithm consists of four distinct phases in each round:

- *Global Model Dissemination:* At the beginning of the t-th round, M edge devices are uniformly selected by the central server to participate in the local training. We denote the set of scheduled edge devices in round t as $\mathcal{M}_t$. Then, the central server disseminates its current global model parameter $\boldsymbol{x}^t$ to the selected edge devices.
- *Local Model Update:* Each selected edge device $i \in \mathcal{M}_t$ uses the received model parameter $\boldsymbol{x}^t$ to initialize its local model $\boldsymbol{x}_i^{(t,0)}$, i.e., $\boldsymbol{x}_i^{(t,0)} = \boldsymbol{x}^t$, and then takes a total of H iterates of stochastic zeroth-order updates. Specifically, at the k-th iteration of the t-th round, edge device computes a stochastic gradient estimator according to (4.2). For the simplicity of notation, we denote the estimated gradient as

$$\boldsymbol{e}_i^{(t,k)} = \widetilde{\nabla} F_i\left(\boldsymbol{x}_i^{(t,k)}, \{\xi_{i,m}^{(t,k)}\}_{m=1}^{b_1}, \{\boldsymbol{v}_{i,n}^{(t,k)}\}_{n=1}^{b_2}, \mu\right), \tag{4.5}$$

Algorithm 6: FedZO algorithm

Input: Initial model $\boldsymbol{x}^0$, learning rate η, step size μ, mini-batch sizes b_1, b_2, number of participating edge devices M **for** $t \in \{0, 1, \ldots, T-1\}$ **do**

  Uniformly sample a subset $\mathcal{M}_t$ of M edge devices
  Disseminate global model $\boldsymbol{x}^t$ to all edge devices in set $\mathcal{M}_t$
  for ***edge device*** $i \in \mathcal{M}_t$ ***in parallel*** **do**
    Initialize local model $\boldsymbol{x}_i^{(t,0)} = \boldsymbol{x}^t$
    for $k = 0, \ldots, H-1$ **do**
      Generate $\{\xi_{i,m}^{(t,k)}\}_{m=1}^{b_1} \sim \mathcal{D}_i$ independently
      Generate $\{\boldsymbol{v}_{i,n}^{(t,k)}\}_{n=1}^{b_2} \sim \mathcal{U}(\mathbb{S}^d)$ independently
      Update $\boldsymbol{x}_i^{(t,k+1)}$ by (4.6)
    end
    Compute local model updates $\boldsymbol{\Delta}_i^t = \boldsymbol{x}_i^{(t,H)} - \boldsymbol{x}_i^{(t,0)}$
    Upload local model updates to central server
  end
  Aggregate local changes $\boldsymbol{\Delta}^t = \frac{1}{M}\sum_{i\in\mathcal{M}_t} \boldsymbol{\Delta}_i^t$
  Update global model $\boldsymbol{x}^{t+1} = \boldsymbol{x}^t + \boldsymbol{\Delta}^t$

end

where $\boldsymbol{x}_i^{(t,k)}$ represents the local model of edge device i at the k-th iteration of the t-th round. Subsequently, the sampled edge devices update their local models as

$$\boldsymbol{x}_i^{(t,k+1)} = \boldsymbol{x}_i^{(t,k)} - \eta \boldsymbol{e}_i^{(t,k)}, k = 0,1,\ldots, H-1, \tag{4.6}$$

where η denotes the learning rate. After H local iterates, edge device i obtains an updated local model $\boldsymbol{x}_i^{(t,H)}$.

- *Local Model Uploading:* In this round, all the selected edge devices calculate the updates of their local models, i.e., $\boldsymbol{\Delta}_i^t = \boldsymbol{x}_i^{(t,H)} - \boldsymbol{x}_i^{(t,0)}$, $i \in \mathcal{M}_t$, and then upload these updates to the central server.
- *Global Model Update:* Once receiving local model updates from the sampled edge devices, the central server aggregates these updates as $\boldsymbol{\Delta}^t = \frac{1}{M}\sum_{i\in\mathcal{M}_t} \boldsymbol{\Delta}_i^t$, and then updates the global model as $\boldsymbol{x}^{t+1} = \boldsymbol{x}^t + \boldsymbol{\Delta}^t$.

4.3 Convergence Analysis for FedZO

This section analyzes the convergence of the FedZO algorithm for both full and partial device participation scenarios. To ensure our analysis caters to more practical situations, we concentrate on settings involving non-convex loss functions and non-i.i.d. data. To facilitate the convergence analysis, the following assumptions are made:

Assumption 4 *The global loss in* (4.1)*, i.e.,* $f(\boldsymbol{x})$*, is lower bounded by* f_**, i.e.,* $f(\boldsymbol{x}) \geq f_* > -\infty$.

Assumption 5 $F_i(\boldsymbol{x}, \xi_i)$*,* $f_i(\boldsymbol{x})$*, and* $f(\boldsymbol{x})$ *are L-smooth. Mathematically, for any* $\boldsymbol{x} \in \mathbb{R}^d$ *and* $\boldsymbol{y} \in \mathbb{R}^d$*, we have*

$$\|\nabla f_i(\boldsymbol{y}) - \nabla f_i(\boldsymbol{x})\| \leq L\|\boldsymbol{y} - \boldsymbol{x}\|, \ \forall i,$$

$$f(\boldsymbol{y}) \leq f(\boldsymbol{x}) + \langle \nabla f(\boldsymbol{x}), \boldsymbol{y} - \boldsymbol{x} \rangle + \frac{L}{2} \|\boldsymbol{y} - \boldsymbol{x}\|^2.$$

Assumption 6 *The second-order moment of stochastic gradient* $\nabla F_i(\boldsymbol{x}, \xi_i)$ *satisfies* $\mathbb{E}_{\xi_i} \|\nabla F_i(\boldsymbol{x}, \xi_i)\|^2 \leq c_g \|\nabla f_i(\boldsymbol{x})\|^2 + \sigma_g^2$*,* $\forall \boldsymbol{x} \in \mathbb{R}^d$*,* $\forall i$*, where* $c_g \geq 1$.

Assumption 7 *The gradient dissimilarity between each local loss function and the global loss function is bounded as* $\|\nabla f(\boldsymbol{x}) - \nabla f_i(\boldsymbol{x})\|^2 \leq c_h \|\nabla f(\boldsymbol{x})\|^2 + \sigma_h^2$*,* $\forall \boldsymbol{x} \in \mathbb{R}^d$*,* $\forall i$*, where* c_h *is a positive constant.*

It is worth noting that the above assumptions are only required for theoretical analysis which are standard in zeroth-order optimization [8, 24].

4.3.1 Full Device Participation

We first characterize the convergence of the FedZO algorithm with full device participation in Theorem 4.1.

Theorem 4.1 *Suppose Assumptions 4–7 hold and the learning rate satisfies*

$$\eta \leq \min \left\{ \frac{N}{72\tilde{c}_g \tilde{c}_h L}, \frac{2}{NH^2L}, \frac{1}{3\sqrt{\tilde{c}_g} HL} \right\}, \tag{4.7}$$

the FedZO algorithm with full device participation satisfies

$$\min_{t \in [T]} \mathbb{E} \left\| \nabla f\left(\boldsymbol{x}^t\right) \right\|^2 \leq 4 \frac{f\left(\boldsymbol{x}^0\right) - f_*}{HT\eta} + \eta \frac{24L}{N} \tilde{\sigma}^2 + \frac{dL^2 \mu^2}{12} + 5L^2 \mu^2, \tag{4.8}$$

where $\tilde{\sigma}^2 = 3\left(1 + \frac{c_g d}{b_1 b_2}\right) \sigma_h^2 + \frac{d\sigma_g^2}{b_1 b_2}$*,* $\tilde{c}_g = 1 + \frac{c_g d}{b_1 b_2}$*, and* $\tilde{c}_h = 1 + c_h$.

Proof For the simplicity of notation, we denote $\delta_t = \mathbb{E}_{\zeta_t} \left[\frac{1}{N} \sum_{i=1}^{N} \sum_{k=0}^{H-1} \|\boldsymbol{x}_i^{(t,k)} - \boldsymbol{x}^t\|^2 \right]$. Before proving Theorem 4.1, we present the following two lemmas. The first lemma characterizes how the global loss, i.e., $f(\boldsymbol{x}^t)$, evolves as the iteration continues.

Lemma 4.1 *With Assumptions 4-7 and full device participation, by letting* $\eta \leq \frac{1}{2HL}$, *we have*

$$\mathbb{E}_t\left[f\left(\boldsymbol{x}^{t+1}\right)\right] \leq f\left(\boldsymbol{x}^t\right) - \left(\frac{\eta H}{2} - \eta^2 \frac{6\tilde{c}_g \tilde{c}_h HL}{N}\right) \left\|\nabla f\left(\boldsymbol{x}^t\right)\right\|^2 + \left(\eta L^2 + \eta^2 \frac{6\tilde{c}_g L^3}{N}\right)\delta_t + \eta^2 \frac{2HL}{N}\tilde{\sigma}^2 + \eta^2 \frac{2HL^3\mu^2}{N} + \eta^2 \frac{d^2 HL^3\mu^2}{2Nb_1b_2} + \eta HL^2\mu^2. \tag{4.9}$$

Please refer to [6] for the proof.

Lemma 4.1 implies that we need to bound δ_t, which is tackled by the following lemma.

Lemma 4.2 *With Assumptions 4-7 and* $\eta \leq \frac{1}{3HL\sqrt{\tilde{c}_g}}$, *we have*

$$\delta_t \leq 3\eta^2 \tilde{c}_g \tilde{c}_h H^3 \left\|\nabla f(\boldsymbol{x}^t)\right\|^2 + H^3L^2\eta^2\mu^2 + \eta^2 H^3 \tilde{\sigma}^2 + \frac{d^2H^3L^2}{4b_1b_2}\eta^2\mu^2.$$

Please refer to [6] for the proof.

By substituting the upper bound of δ_t in Lemma 4.2 into (4.9), we obtain

$$\mathbb{E}_t\left[f\left(\boldsymbol{x}^{t+1}\right)\right] \leq f\left(\boldsymbol{x}^t\right) - \left(\frac{\eta H}{2} - 3\eta^2 \tilde{c}_g \tilde{c}_h Q_1(\eta)\right)\left\|\nabla f(\boldsymbol{x}^t)\right\|^2 + \eta^2\tilde{\sigma}^2 Q_1(\eta) + Q_1(\eta)\left(L^2\eta^2\mu^2 + \frac{d^2L^2}{4b_1b_2}\eta^2\mu^2\right) + \eta HL^2\mu^2, \tag{4.10}$$

where $Q_1(\eta) = \frac{2HL}{N} + (\eta L^2 + \eta^2 \frac{6\tilde{c}_g L^3}{N})H^3$. Under condition (4.7), we have

$$Q_1(\eta) \leq \frac{6HL}{N}, \ \eta^2 \frac{18\tilde{c}_g \tilde{c}_h HL}{N} \leq \frac{\eta H}{4}, \ L^2\eta^2\mu^2 \leq \frac{NL\eta\mu^2}{72},$$

and $\frac{d^2L^2\eta^2\mu^2}{4b_1b_2} \leq \frac{d^2NL\eta\mu^2}{288b_1b_2\tilde{c}_g\tilde{c}_h}$. Recalling the definition of $\tilde{c}_g$ and the fact that $c_g\tilde{c}_h \geq 1$, we have $\frac{d^2}{b_1b_2\tilde{c}_g\tilde{c}_h} \leq \frac{d}{c_g\tilde{c}_h} \leq d$. With the above inequalities, we have

$$\mathbb{E}_t\left[f\left(\boldsymbol{x}^{t+1}\right)\right] \leq f\left(\boldsymbol{x}^t\right) - \frac{\eta H}{4}\left\|\nabla f\left(\boldsymbol{x}^t\right)\right\|^2 + \eta^2 \frac{6HL}{N}\tilde{\sigma}^2 + \frac{\eta dHL^2\mu^2}{48} + \frac{13}{12}\eta HL^2\mu^2. \tag{4.11}$$

By taking expectation on both sides of (4.11) and telescoping from $t = 0$ to $T - 1$, we obtain

$$\frac{1}{T}\sum_{t=0}^{T-1}\mathbb{E}\left\|\nabla f\left(\boldsymbol{x}^t\right)\right\|^2 \leq 4\frac{f\left(\boldsymbol{x}^0\right)-\mathbb{E}\left[f\left(\boldsymbol{x}^T\right)\right]}{HT\eta}+\eta\frac{24L}{N}\tilde{\sigma}^2+\frac{dL^2\mu^2}{12}+5L^2\mu^2. \tag{4.12}$$

With Assumption 4, i.e., $f(\boldsymbol{x}) \geq f_*$, we obtain Theorem 4.1. □

The subsequent corollary can be derived by appropriately substituting a learning rate η and a step size μ into Theorem 4.1.

Corollary 4.1 *Suppose Assumptions 4–7 hold and let $b_1b_2 \leq d$, $\mu = (db_1b_2NH T)^{-\frac{1}{4}}$, and $\eta = (Nb_1b_2)^{\frac{1}{2}}(dHT)^{-\frac{1}{2}}$, which holds for* (4.7) *if T is large enough. The FedZO algorithm with full device participation satisfies*

$$\min_{t\in[T]}\mathbb{E}\left\|\nabla f\left(\boldsymbol{x}^t\right)\right\|^2 \leq O\left(d^{\frac{1}{2}}(NHTb_1b_2)^{-\frac{1}{2}}\right)+O\left((db_1b_2NHT)^{-\frac{1}{2}}\right), \tag{4.13}$$

where the right hand side of (4.13) *is dominated by $O\left(d^{\frac{1}{2}}(NHTb_1b_2)^{-\frac{1}{2}}\right)$.*

In light of Corollary 4.1, it can be observed that the FedZO algorithm is capable of achieving a convergence rate of $O\left(d^{\frac{1}{2}}(NHTb_1b_2)^{-\frac{1}{2}}\right)$. Notably, FedZO demonstrates a linear speedup with respect to the quantity of local iterations and the number of participating edge devices in comparison to the centralized zeroth-order algorithm (i.e., ZO-SGD) that attains a convergence rate of $O\left(d^{\frac{1}{2}}T^{-\frac{1}{2}}\right)$ [9]. To provide a fair comparison, $b_1 = b_2 = 1$ is assumed.

To achieve equivalent accuracy, the number of communication rounds necessitated by the FedZO algorithm can be diminished by a factor of H when compared with DZOPA, which has a convergence rate of $O\left(d^{\frac{1}{2}}(NT)^{-\frac{1}{2}}\right)$ [24]. Notably, the convergence rate of the FedZO algorithm is contingent upon the dimension of the model parameter. Specifically, the convergence rate of FedZO is $\sqrt{d}$ times more gradual than that of its first-order analogue, namely FedAvg. This deterioration is consistent with its centralized counterpart [9].

4.3.2 *Partial Device Participation*

We characterize the convergence of the FedZO algorithm with partial device participation in the following theorem.

Theorem 4.2 *Suppose Assumptions 4-7 hold and the learning rate satisfies*

$$\eta \leq \min\left\{\frac{M}{192\tilde{c}_g\tilde{c}_h L}, \frac{M}{72c_h HL}, \frac{2}{MH^2L}, \frac{1}{3\sqrt{\tilde{c}_g}HL}, \frac{1}{3\sqrt{MH^3L}}\right\}, \quad (4.14)$$

the FedZO algorithm with partial device participation satisfies

$$\min_{t\in[T]} \mathbb{E}\left\|\nabla f\left(\boldsymbol{x}^t\right)\right\|^2 \leq 4\frac{f\left(\boldsymbol{x}^0\right)-f_*}{HT\eta}+\eta\frac{32L}{M}\tilde{\sigma}^2$$
$$+\eta\frac{36HL\sigma_h^2}{M}+\frac{dL^2\mu^2}{24}+13L^2\mu^2, \quad (4.15)$$

where $\tilde{c}_g$, $\tilde{c}_h$, and $\tilde{\sigma}^2$ are defined in Theorem 4.1.

Proof We first characterize how the global loss, i.e., $f(\boldsymbol{x}^t)$, evolves as the iteration continues in the following lemma.

Lemma 4.3 *With Assumptions 4–7 and partial device participation, by letting the learning rate $\eta \leq \frac{1}{2HL}$, we have*

$$\mathbb{E}_t\left[f\left(\boldsymbol{x}^{t+1}\right)\right] \leq f\left(\boldsymbol{x}^t\right)-\left(\frac{\eta H}{2}-\eta^2\frac{6\tilde{c}_g\tilde{c}_h HL}{M}-\eta^2\frac{9H^2Lc_h}{M}\right)\left\|\nabla f\left(\boldsymbol{x}^t\right)\right\|^2$$
$$+\left(\eta L^2+\eta^2\frac{6\tilde{c}_g L^3}{M}+\eta^2 18HL^3\right)\delta_t+\eta^2\frac{2HL}{M}\tilde{\sigma}^2+\frac{9\eta^2H^2L\sigma_h^2}{M}$$
$$+\eta^2\frac{2HL^3\mu^2}{M}+\eta^2\frac{d^2HL^3\mu^2}{2Mb_1b_2}+6\eta^2H^2L^3\mu^2+\eta HL^2\mu^2.$$

Please refer to [6] for the proof.

By combining Lemmas 4.2 and 4.3, we have

$$\mathbb{E}_t\left[f\left(\boldsymbol{x}^{t+1}\right)\right] \leq f\left(\boldsymbol{x}^t\right)-\left(\frac{\eta H}{2}-\eta^2\frac{9H^2Lc_h}{M}-3\tilde{c}_g\tilde{c}_h\eta^2Q_2(\eta)\right)\left\|\nabla f\left(\boldsymbol{x}^t\right)\right\|^2$$
$$+\eta^2\tilde{\sigma}^2Q_2(\eta)+\frac{9\eta^2H^2L\sigma_h^2}{M}+Q_2(\eta)\left(L^2\eta^2\mu^2+\frac{d^2L^2}{4b_1b_2}\eta^2\mu^2\right)$$
$$+6\eta^2H^2L^3\mu^2+\eta HL^2\mu^2,$$

where $Q_2(\eta) = \frac{2HL}{M} + \left(\eta L^2 + \eta^2 \frac{6\tilde{c}_g L^3}{M} + \eta^2 18HL^3\right) H^3$. Under condition (4.14), we have

$$\eta^2 \frac{9H^2 L c_h}{M} \le \frac{\eta H}{8},\ Q_2(\eta) \le \frac{8HL}{M},\ \frac{24\eta^2 \tilde{c}_g \tilde{c}_h HL}{M} \le \frac{\eta H}{8},$$

$$L^2 \eta^2 \mu^2 \le \frac{ML\eta\mu^2}{192},\ \frac{d^2 L^2 \eta^2 \mu^2}{4b_1 b_2} \le \frac{d^2 ML\eta\mu^2}{768 b_1 b_2 \tilde{c}_g \tilde{c}_h},$$

and $6\eta^2 H^2 L^3 \mu^2 \le 2\eta H L^2 \mu^2$. Recalling the definition of $\tilde{c}_g$ and the fact that $c_g \tilde{c}_h \ge 1$, we have $\frac{d^2}{b_1 b_2 \tilde{c}_g \tilde{c}_h} \le \frac{d}{c_g \tilde{c}_h} \le d$. With the above inequalities, we have

$$\mathbb{E}_t\left[f\left(\boldsymbol{x}^{t+1}\right)\right] \le f\left(\boldsymbol{x}^t\right) - \frac{\eta H}{4}\left\|\nabla f\left(\boldsymbol{x}^t\right)\right\|^2 + \eta^2 \frac{8HL}{M}\tilde{\sigma}^2 + \frac{9\eta^2 H^2 L \sigma_h^2}{M}$$
$$+ \frac{\eta H d L^2 \mu^2}{96} + \left(3 + \frac{1}{24}\right)\eta H L^2 \mu^2.$$

By following the similar steps as in proving Theorem 4.1, we obtain Theorem 4.2. □

The third term presented in (4.15) is absent from (4.8). This discrepancy can be attributed to the stochasticity introduced by device sampling, whereas the complete participation of devices eradicates such randomness, consequently diminishing the optimality gap.

Similarly, we can deduce the subsequent corollary by appropriately substituting an optimal learning rate η and an ideal step size μ into Theorem 4.2.

Corollary 4.2 *Suppose Assumptions 4–7 hold and let $b_1 b_2 \le d$, $\mu = (d b_1 b_2 M HT)^{-\frac{1}{4}}$, and $\eta = (M b_1 b_2)^{\frac{1}{2}} (dHT)^{-\frac{1}{2}}$, which holds for* (4.14) *if T is large enough. The FedZO algorithm with partial device participation satisfies*

$$\min_{t \in [T]} \mathbb{E}\left\|\nabla f\left(\boldsymbol{x}^t\right)\right\|^2 \le O\left(d^{\frac{1}{2}} (MHT b_1 b_2)^{-\frac{1}{2}}\right)$$
$$+ O\left((b_1 b_2 H)^{\frac{1}{2}} (dMT)^{-\frac{1}{2}}\right) + O\left((d b_1 b_2 MHT)^{-\frac{1}{2}}\right). \tag{4.16}$$

Obviously, the number of local iterates cannot exceed $O(d(b_1 b_2)^{-1}))$ to attain a linear speedup in terms of the number of local iterates and participating edge devices. Combining it with constraint (4.14), the largest value of H is obtained as

$$\min\left\{O\left((dT)^{\frac{1}{3}} (b_1 b_2)^{-\frac{1}{3}} M^{-1}\right), O\left(TM^{-1}\right), O\left(d (b_1 b_2)^{-1}\right)\right\}.$$

4.4 AirComp-Assisted FedZO Algorithm

We next introduce the implementation of the FedZO algorithm over wireless networks using AirComp [4, 7, 12]. Within each communication round, both the downlink model dissemination and the uplink model uploading are realized via wireless fading channels. Since the transmit power of the central server is significantly higher that of edge devices, the downlink model dissemination is presumed to be error-free, directing our attention toward the uplink model uploading.

4.4.1 Over-the-Air Model Aggregation

Consider a wireless FEEL system in which the server and all edge devices deploy a single antenna. Over wireless fading channels, the transmission of local model may suffer from adverse channel distortion, which results in the degradation of convergence performance of the AirComp-assisted FedZO algorithm. To deal with this problem, we set a threshold $h_{\min}$ and choose a subset of edge devices $\mathcal{M}_t = \{i \mid |h_i^t| \geq h_{\min}\}$ to participate in the training, where $h_i^t \in \mathbb{C}$ denotes the channel coefficient between edge device i and the central server in round t. It is assumed that h_i^t are i.i.d. across distinct edge devices and communication rounds [18, 19]. Notably, the applied device scheduling strategy can be regarded as uniform sampling, as analyzed in Chap. 4.3.2.

The scheduled edge devices simultaneously transmit their precoded model updates, denoted by $\alpha_i^t \boldsymbol{\Delta}_i^t$, to the central server, where α_i^t represents the transmit scalar of edge device i during the t-th round. Notably, the synchronization among distributed edge devices can be realized via various methods proposed in [1, 14, 18, 19, 23]. The model update vector $\boldsymbol{\Delta}_i^t$, with dimension d, is assumed to be transmitted within a single transmission block, while the channel coefficient is assume to remain constant during one transmission block [1, 14, 18, 19, 23]. Consequently, we can express the aggregated signal received at the central server as

$$\boldsymbol{s}^t = \sum_{i \in \mathcal{M}_t} h_i^t \alpha_i^t \boldsymbol{\Delta}_i^t + \boldsymbol{n}_t, \tag{4.17}$$

where $\boldsymbol{n}_t \sim \mathcal{CN}(0, \sigma_w^2 \boldsymbol{I}_d)$ denotes the AWGN vector at the central server.

4.4.2 Transceiver Design

The transmitted signal at each edge device is subject to an energy constraint during one communication round, i.e., $\left\|\alpha_i^t \boldsymbol{\Delta}_i^t\right\|^2 \leq dP$, where dP is the total energy of each edge device in one communication round. We assume that the channel state information (CSI) is available at both the central server and edge devices as in [1, 14, 18, 19, 23]. To meet the energy constraint of each edge device, the transmit scalar of device i is set as

$$\alpha_i^t = \frac{h_{\min}}{h_i^t}\sqrt{\frac{dP}{\Delta_{\max}^t}}, \ \forall i, \tag{4.18}$$

where $\Delta_{\max}^t = \max_{i\in\mathcal{M}_t} \left\|\boldsymbol{\Delta}_i^t\right\|^2$. Thus, the received signal is given by

$$\boldsymbol{s}^t = \sqrt{\frac{dPh_{\min}^2}{\Delta_{\max}^t}} \sum_{i\in\mathcal{M}_t} \boldsymbol{\Delta}_i^t + \boldsymbol{n}_t. \tag{4.19}$$

To recover the desired global model update

$$\boldsymbol{\Delta}^t = \frac{1}{|\mathcal{M}_t|}\sum_{i\in\mathcal{M}_t} \boldsymbol{\Delta}_i^t, \tag{4.20}$$

where $|\mathcal{M}_t|$ denotes cardinality of set $\mathcal{M}_t$, the central server scales $\boldsymbol{s}^t$ with a receive scalar $\frac{1}{|\mathcal{M}_t|}\sqrt{\frac{\Delta_{\max}^t}{dPh_{\min}^2}}$, and obtains a noisy version of the global model update as follows

$$\boldsymbol{y}^t = \boldsymbol{\Delta}^t + \tilde{\boldsymbol{n}}_t, \tag{4.21}$$

where $\tilde{\boldsymbol{n}}_t \sim \mathcal{CN}\left(0, \frac{\sigma_w^2\Delta_{\max}^t}{|\mathcal{M}_t|^2 dPh_{\min}^2}\boldsymbol{I}_d\right)$. As a result, the global model at the central server is updated as $\boldsymbol{x}^{t+1} = \boldsymbol{x}^t + \boldsymbol{\Delta}^t + \tilde{\boldsymbol{n}}_t$. To facilitate the transceiver design, each edge device needs to know the maximum of squared norm of local model updates among the participating edge devices, i.e., $\Delta_{\max}^t$, and the instantaneous channel coefficient between itself and the central server, i.e., h_i^t, which can be obtained via feedback from the central server. Before uplink model aggregation, the central server collects the squared norm of local model update $\left\|\boldsymbol{\Delta}_i^t\right\|^2$ from each edge device $i \in \mathcal{M}_t$, and then broadcasts $\Delta_{\max}^t$ to all edge devices. Besides, the central server estimates and feeds back the channel coefficients to these corresponding edge devices. It is worth noting that the communication overhead introduced by the exchange of these scalars is negligible when compared with the transmission of high-dimensional model parameters.

4.4.3 Convergence Analysis for AirComp-Assisted FedZO

The following theorem characterizes the convergence of the AirComp-assisted FedZO algorithm.

Theorem 4.3 *Suppose Assumptions 4–7 hold and the learning rate satisfies*

$$\eta \le \min\left\{\frac{\tilde{M}}{288\tilde{c}_g\tilde{c}_h L}, \frac{\tilde{M}}{108c_h HL}, \frac{3}{2NH^2L}, \frac{1}{3\sqrt{\tilde{c}_g}HL},\right.$$
$$\left.\frac{1}{2\sqrt{3NH^3}L}, \frac{\sqrt{\tilde{M}\gamma}}{L\sqrt{2\tilde{c}_g NH^3}}, \frac{\tilde{M}^2\gamma}{36\tilde{c}_g\tilde{c}_h NHL}\right\}, \tag{4.22}$$

where $\tilde{M} = \min\{|\mathcal{M}_t|, t \in [T]\}$. *The AirComp-assisted FedZO algorithm satisfies*

$$\min_{t\in[T]} \mathbb{E}\left\|\nabla f\left(\boldsymbol{x}^t\right)\right\|^2 \le 4\frac{f\left(\boldsymbol{x}^0\right) - f_*}{HT\eta} + \eta\frac{32L}{\tilde{M}}\hat{C}\tilde{\sigma}^2$$
$$+\eta\frac{36HL\sigma_h^2}{\tilde{M}} + \hat{C}\frac{dL^2\mu^2}{36} + \left(12 + \frac{\hat{C}}{9}\right)L^2\mu^2, \tag{4.23}$$

where $\hat{C} = 1 + \frac{NH}{8\tilde{M}\gamma}$ *and* $\gamma = \frac{Ph_{\min}^2}{\sigma_w^2}$. $\tilde{c}_g$, $\tilde{c}_h$, *and* $\tilde{\sigma}^2$ *are defined in Theorem 4.1.*

Theorem 4.3 shows that the convergence rate depends on γ, which is the minimum receive SNR. When γ goes to infinity, i.e., noise-free case, Theorem 4.3 almost reduces to Theorem 4.2. Obviously, a smaller value of SNR leads to a slower convergence speed. The following corollary illustrates that a same-order convergence rate as the noise-free case presented in Sect. 4.3.2 can be attained.

4.5 Simulation

In this section, we conduct extensive simulations to evaluate the effectiveness of the FedZO algorithm for the application of federated black-box attack.

4.5.1 Simulation Settings

The goal of federated black-box attacks is to collaboratively generate a common perturbation such that the perturbed images are visually imperceptible to a human but could mislead the classifier. The attack loss of image z_i is given by

$$\begin{aligned}\psi_i(\boldsymbol{x}) = \max\Bigg\{&\Phi_{y_i}\left(\frac{1}{2}\tanh\left(\tanh^{-1}2z_i+\boldsymbol{x}\right)\right)\\&-\max_{j\neq y_i}\left\{\Phi_j\left(\frac{1}{2}\tanh\left(\tanh^{-1}2z_i+\boldsymbol{x}\right)\right)\right\},0\Bigg\}\\&+c\left\|\frac{1}{2}\tanh\left(\tanh^{-1}2z_i+\boldsymbol{x}\right)-z_i\right\|^2,\end{aligned} \tag{4.24}$$

where y_i denotes the label of image z_i, $\Phi_j(z)$ represents the prediction confidence of image z to class j, $\frac{1}{2}\tanh\left(\tanh^{-1}2z_i+\boldsymbol{x}\right)$ is the adversarial example of z_i, $\frac{1}{2}\tanh\left(\tanh^{-1}2z_i+\boldsymbol{x}\right)-z_i$ is the distortion perturbed by $\boldsymbol{x}$ in the original image space. Parameter c balances the trade-off between the adversarial success and distortion loss. The attack loss of device n can be expressed as $f_n(\boldsymbol{x}) = \frac{1}{|\mathcal{D}_n|}\sum_{i\in\mathcal{D}_n}\psi_i(x)$, where $\mathcal{D}_n$ denotes the dataset at edge device n. Mathematically, the problem of federated black-box attack of a DNN model is $\min_{\boldsymbol{x}\in\mathbb{R}^d}\frac{1}{N}\sum_{n=1}^{N}f_n(\boldsymbol{x})$. Note that the inner structure of the DNN model is not available, we thus do not have the analytic information of $\Phi_j(z)$, which makes the federated black-box attack problem lies in the category of zeroth-order optimization.

In this experiment setting, all edge devices share one well-trained DNN classifier[1] that has a testing accuracy of 82.3% on CIFAR-10 dataset [2]. We pick 4992 correctly classified samples from the training set of image class "deer" (containing 5000 samples) and then distribute these samples to edge devices without overlapping. Each edge device is assigned a random number of samples. We set the balancing parameter $c = 1$. The mini-batch sizes are set to $b_1 = 25$ and $b_2 = 20$. The learning rate and step size are set to $\eta = 0.001$ and $\mu = 0.001$, respectively.

4.5.2 Simulation Results

Figure 4.1 shows the impact of the number of local updates on the convergence performance of the FedZO algorithm with full device participation. Specifically, we vary the number of local updates $H \in \{5, 10, 20, 50\}$ and present the attack loss versus the number of communication rounds. We can observe that the FedZO algorithm can effectively reduce the attack loss for different values of H. Besides, the convergence speed of the FedZO algorithm tends to increase with the increase of H. This demonstrates the speedup in the number of the local iterates as shown in Sect. 4.3. We further compare the performance of the FedZO algorithm with DZOPA [24] and ZONE-S [11]. For DZOPA, the learning rate (i.e., η) and step size (i.e., μ)

[1] https://github.com/carlini/nn_robust_attacks.

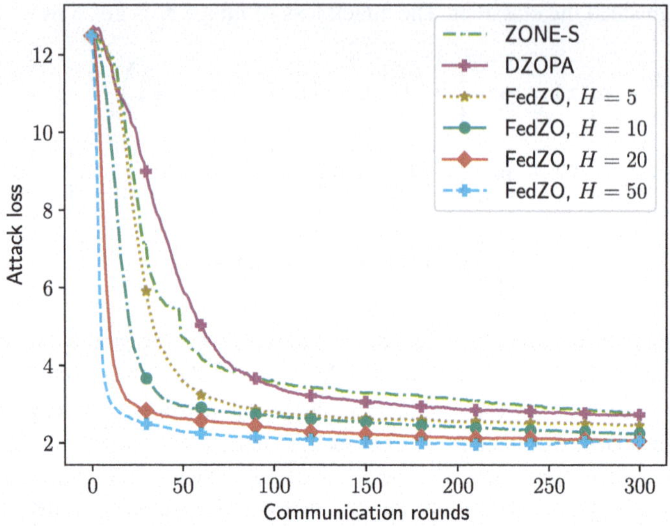

Fig. 4.1 Impact of number of local updates on attack loss

are set to 0.005 and 0.001, respectively. For ZONE-S, the penalty parameter (i.e., ρ, defined in [11]) and step size (i.e., μ) are set to 500 and 0.001, respectively. Notably, DZOPA is proposed for the peer-to-peer architecture and cannot be directly applied to our considered server-client architecture. For a fair comparison, we depict the performance of DZOPA under a fully-connected graph and upgrade the two-point stochastic gradient estimator of [24] to a mini-batch-type one as in (4.2). Results show that the FedZO algorithm outperforms the baselines even when $H = 5$. With a larger number of local updates, the attack loss of the FedZO algorithm decreases much faster than that of the baselines.

In Fig. 4.2, we show the convergence performance of the FedZO algorithm with different number of participating edge devices and fixed $H = 20$. The number of participating edge devices M varies over set $\{5, 10, 25, 50\}$. It is evident that the FedZO algorithm can realize a small attack loss under different values of M. Besides, it can be observed that a larger number of edge devices leads to a better convergence speed. We can observe the speedup in the number of participating devices, which matches well with our analysis in Sect. 4.3.

Figure 4.3 shows the performance of the AirComp-assisted FedZO algorithm presented in Sect. 4.4. Without loss of generality, we model the channel as $h_i^t \sim \mathcal{CN}(0, 1)$, $\forall i, t$, and set the threshold $h_{\min}$ to 0.8. Besides, we set the number of local iterates to $H = 20$. We take the FedZO algorithm with noise-free aggregation as benchmark with $H = 20$, where the participating edge devices are the same as that of the case with noise. We investigate the attack loss versus the number of communication rounds under different SNR, i.e., $P/\sigma_w^2 \in \{-10\,\text{dB},\ -5\,\text{dB},\ 0\,\text{dB}\}$. As can be observed, the convergence of the FedZO algorithm is well preserved under such SNR settings. Besides, increasing the SNR accelerates the convergence speed

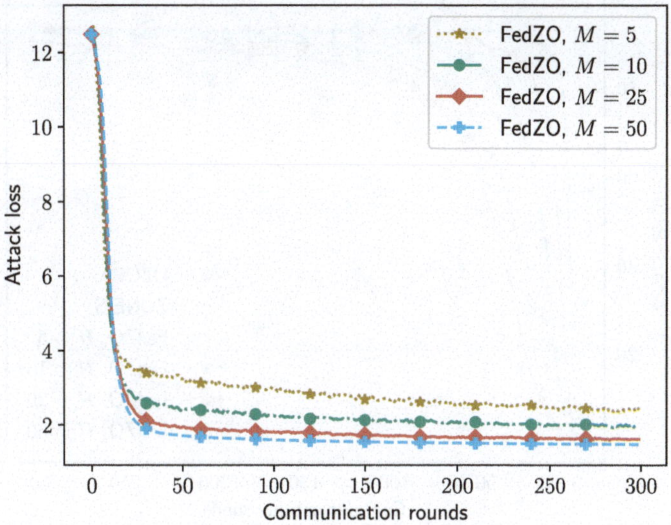

Fig. 4.2 Impact of number of participating edge devices on attack loss

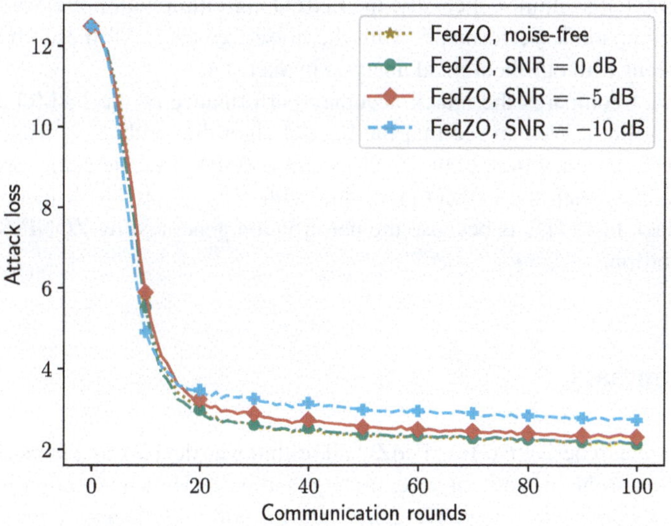

Fig. 4.3 Impact of SNR on the performance of the AirComp-assisted FedZO algorithm

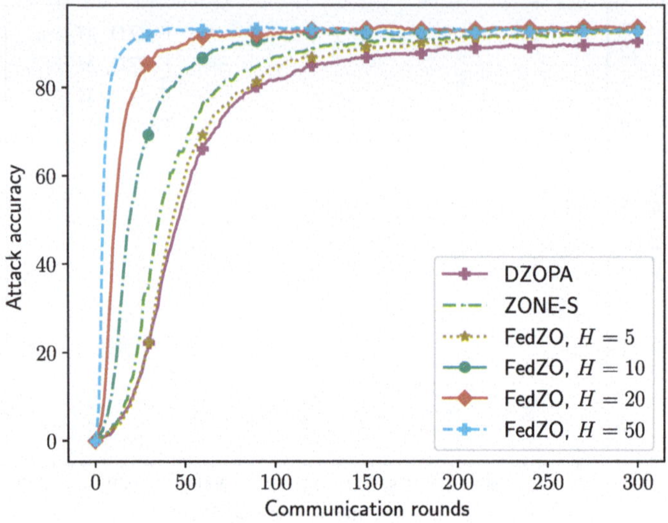

Fig. 4.4 Impact of number of local updates on attack accuracy

of the FedZO algorithm. Especially, the FedZO algorithm with noise SNR $= 0$ dB attains a comparable performance with the noise-free case. All these observations are consistent with our theoretical analysis in Sect. 4.4.3.

Figure 4.4 compares the attack accuracy performance of the FedZO algorithm and the baselines. We can see that the FedZO algorithm with $H \geq 10$ achieves a better attack accuracy than both ZONE-S and DZOPA. Despite achieving a better attack accuracy than the FedZO algorithm with $H = 5$, ZONE-S incurs a much higher attack loss. This is because the perturbation generated by ZONE-S brings a large distortion.

4.6 Summary

In this chapter, a derivative-free FedZO algorithm was devised to address federated optimization problems without necessitating gradient or Hessian information. Under non-convex conditions, the convergence rate for non-i.i.d. data was characterized, and a linear speedup concerning the number of participating devices and local iterations was exhibited. Following this, the convergence assurance for the AirComp-assisted FedZO algorithm was established, thereby facilitating the execution of the algorithm over wireless networks. Simulation results illustrated the efficiency of the FedZO algorithm. It was also corroborated that the AirComp-assisted FedZO algorithm could achieve a performance comparable to that of the noise-free scenario under specific SNR regime.

References

1. X. Cao, G. Zhu, J. Xu, Z. Wang, and S. Cui. Optimized power control design for over-the-air federated edge learning. *IEEE J. Sel. Areas Commun.*, pages 1–1, 2021, doi:10.1109/JSAC.2021.3126060.
2. N. Carlini and D. Wagner. Towards evaluating the robustness of neural networks. In *Proc. Symp. Security Privacy (SP)*, 2017.
3. Z. Dai, B. K. H. Low, and P. Jaillet. Federated bayesian optimization via thompson sampling. In *Proc. Neural Inf. Process. Syst. (NeurIPS)*, 2020.
4. W. Fang, Y. Jiang, Y. Shi, Y. Zhou, W. Chen, and K. B. Letaief. Over-the-air computation via reconfigurable intelligent surface. *IEEE Trans. Commun.*, 69(12):8612–8626, Dec. 2021.
5. W. Fang, Z. Yu, Y. Jiang, Y. Shi, C. N. Jones, and Y. Zhou. Communication-efficient stochastic zeroth-order optimization for federated learning. *IEEE Trans. Signal Process.*, 70:5058–5073, 2022.
6. W. Fang, Z. Yu, Y. Jiang, Y. Shi, C. N. Jones, and Y. Zhou. Communication-efficient stochastic zeroth-order optimization for federated learning. *IEEE Transactions on Signal Processing*, 70:5058–5073, 2022.
7. W. Fang, Y. Zou, H. Zhu, Y. Shi, and Y. Zhou. Optimal receive beamforming for over-the-air computation. In *2021 IEEE 22nd International Workshop on Signal Processing Advances in Wireless Communications (SPAWC)*, pages 61–65. IEEE, 2021.
8. X. Gao, B. Jiang, and S. Zhang. On the information-adaptive variants of the ADMM: An iteration complexity perspective. *J. Sci. Comput.*, 76(1):327–363, 2018.
9. S. Ghadimi and G. Lan. Stochastic first- and zeroth-order methods for nonconvex stochastic programming. *SIAM J. Optim.*, 23(4):2341–2368, 2013.
10. B. Gu, X. Wei, S. Gao, Z. Xiong, C. Deng, and H. Huang. Black-box reductions for zeroth-order gradient algorithms to achieve lower query complexity. *J. Mach. Learn. Res.*, 22(170):1–47, 2021.
11. D. Hajinezhad, M. Hong, and A. Garcia. ZONE: Zeroth-order nonconvex multiagent optimization over networks. *IEEE Trans. Automat. Contr.*, 64(10):3995–4010, Oct. 2019.
12. K. B. Letaief, Y. Shi, J. Lu, and J. Lu. Edge artificial intelligence for 6G: Vision, enabling technologies, and applications. *IEEE J. Sel. Areas Commun.*, pages 1–1, 2021, doi:10.1109/JSAC.2021.3126076.
13. T. Li, A. K. Sahu, M. Zaheer, M. Sanjabi, A. Talwalkar, and V. Smithy. FedDANE: A federated newton-type method. In *Proc. IEEE Asilomar Conf. Signals, Systems, Computers (ACSSC)*, 2019.
14. H. Liu, X. Yuan, and Y.-J. A. Zhang. Reconfigurable intelligent surface enabled federated learning: A unified communication-learning design approach. *IEEE Trans. Wireless Commun.*, 20(11):7595–7609, Nov. 2021.
15. S. Liu, J. Chen, P.-Y. Chen, and A. Hero. Zeroth-order online alternating direction method of multipliers: Convergence analysis and applications. In *Proc. Int. Conf. Artif. Intell. Stat. (AISTATS)*, 2018.
16. S. Liu, P.-Y. Chen, B. Kailkhura, G. Zhang, A. O. Hero III, and P. K. Varshney. A primer on zeroth-order optimization in signal processing and machine learning: Principals, recent advances, and applications. *IEEE Signal Process. Mag.*, 37(5):43–54, Sept. 2020.
17. B. McMahan, E. Moore, D. Ramage, S. Hampson, and B. A. y. Arcas. Communication-Efficient Learning of Deep Networks from Decentralized Data. In *Proc. Int. Conf. Artif. Intell. Stat. (AISTATS)*, 2017.
18. R. Paul, Y. Friedman, and K. Cohen. Accelerated gradient descent learning over multiple access fading channels. *IEEE J. Sel. Areas Commun.*, pages 1–1, 2021, doi:10.1109/JSAC.2021.3118410.
19. T. Sery, N. Shlezinger, K. Cohen, and Y. C. Eldar. Over-the-air federated learning from heterogeneous data. *IEEE Trans. Signal Process.*, 69:3796–3811, Jun. 2021.

20. Y. Tang, J. Zhang, and N. Li. Distributed zero-order algorithms for nonconvex multiagent optimization. *IEEE Trans. Control. Netw. Syst.*, 8(1):269–281, Mar. 2021.
21. J. Wang, Z. Charles, Z. Xu, G. Joshi, H. B. McMahan, M. Al-Shedivat, G. Andrew, S. Avestimehr, K. Daly, D. Data, et al. A field guide to federated optimization. *arXiv preprint arXiv:2107.06917*, 2021.
22. J. Wang, Q. Liu, H. Liang, G. Joshi, and H. V. Poor. A novel framework for the analysis and design of heterogeneous federated learning. *IEEE Trans. Signal Process.*, 69:5234–5249, 2021.
23. Z. Wang, J. Qiu, Y. Zhou, Y. Shi, L. Fu, W. Chen, and K. B. Letaief. Federated learning via intelligent reflecting surface. *IEEE Trans. Wireless Commun.*, pages 1–1, 2021, doi:10.1109/TWC.2021.3099505.
24. X. Yi, S. Zhang, T. Yang, T. Chai, and K. H. Johansson. Zeroth-order algorithms for stochastic distributed nonconvex optimization. *arXiv preprint arXiv:2106.02958*, 2021.
25. X. Zhang, M. Hong, S. Dhople, W. Yin, and Y. Liu. FedPD: A federated learning framework with adaptivity to non-IID data. *IEEE Trans. Signal Process.*, 69:6055–6070, Oct. 2021.

Part III
Architecture

In this part, we shall exploit various emerging communication and networking techniques (i.e., over-the-air computation, reconfigurable intelligent surface, unmanned aerial vehicle) to tackle the critical challenges of FEEL. The multi-cell architecture presents a feasible solution for collaborative FEEL training among multiple cells.

Chapter 5
GNN for Optimizing RIS-Assisted Federated Edge Learning

5.1 Overview

Benefiting from the inherent signal superposition property, integrating AirComp and FEEL can enable spectrum-efficient uplink model aggregation. However, the performance of AirComp-based FEEL is limited by the worst channel condition towards the edge server. To address this issue, RIS can be deployed to dynamically alter the propagation environment [12, 18, 21, 22]. Specifically, RIS is composed of numerous reflection elements that are passive and can be reconfigured to reflect incoming signals, leading to a well-aligned signal at the receiver [5]. RIS has found its application in many scenarios, including privacy preservation [22], robustness enhancement [10], energy consumption optimization [7, 9], and spectrum efficiency improvement [4, 8]. This chapter presents an over-the-air FEEL strategy involving RIS assistance, where AirComp and RIS work together to achieve prompt and precise uplink model aggregation. We aim to enhance the performance of FEEL via the joint optimization of the AirComp transceiver and RIS phase-shifts.

It poses great challenges to accomplish such a task. First, to characterize the performance of FEEL in terms of communication parameters, the metric can be only obtained by conducting rigorous convergence analysis for the specifically designed wireless FEEL system. Second, the performance of FEEL should be evaluated from a long-term perspective, since it is determined by the communication errors accumulated over all communication rounds. Third, this task involves the joint optimization of the AirComp transceiver and RIS phase-shifts. Alternating optimization algorithm can be developed to solve such a problem, which, however, is computationally expensive.

To address these challenges, we take the aggregation errors accumulated over all communication rounds into consideration, and derive a convergence upper bound of the time-average norm of the global gradient for RIS-assisted over-the-air FEEL. To minimize the time-average error for enhancing the system performance, we propose an effective algorithm based on alternately optimizing the transmit power,

Y. Zhou et al., *Federated Edge Learning*, Wireless Networks,
https://doi.org/10.1007/978-3-031-96649-1_5

denoising factor, and RIS phase-shifts. To reduce the computational complexity and improve the algorithmic scalability [2, 3, 15, 16, 19], we develop a GNN-based learning algorithm. Moreover, we conduct extensive simulations to investigate the performance of the proposed optimization-based and GNN-based algorithms.

5.2 System Model

5.2.1 Learning Model

Consider an RIS-assisted over-the-air FEEL network as depicted in Fig. 5.1, where a edge server coordinates a set $\mathcal{K} = \{1, \ldots, K\}$ of K devices to train a global model with the assistance of an RIS. The edge server and all edge devices are assumed to be equipped with a single antenna. Each edge device $k \in \mathcal{K}$ owns a local dataset $\mathcal{D}_k = \{(\boldsymbol{x}_{km}, y_{km}) \mid 1 \leq m \leq M_k\}$ for local model training, where $(\boldsymbol{x}_{km}, y_{km})$ denotes the m-th input feature and label pair at device k, and M_k denotes the number of training samples available at device k. The training datasets are assumed to be independent and identically distributed (i.i.d.) across different edge devices, and have the same number of training samples, i.e., $M_k = M_j, \forall\, k, j \in \mathcal{K}$. The objective is to find the optimal model parameter $\boldsymbol{w}^* \in \mathbb{R}^{\Omega}$ that minimizes the global loss function $F(\boldsymbol{w})$, i.e., $\min_{\boldsymbol{w}} F(\boldsymbol{w}) = \frac{1}{M} \sum_{k\in\mathcal{K}} M_k F_k(\boldsymbol{w}) = \frac{1}{K} \sum_{k\in\mathcal{K}} F_k(\boldsymbol{w})$, where $M = \sum_{k\in\mathcal{K}} M_k$ denotes the total number of training samples, and $F_k(\boldsymbol{w})$ denotes local loss function at device k. In each round $t = 1, \ldots, T$, the following three steps are performed.

- *Global model dissemination*: At the beginning of round t, each edge device receives global model $\boldsymbol{w}(t-1)$, which is transmitted via the downlink channel from the edge server. Since the transmit power of the edge server is much greater

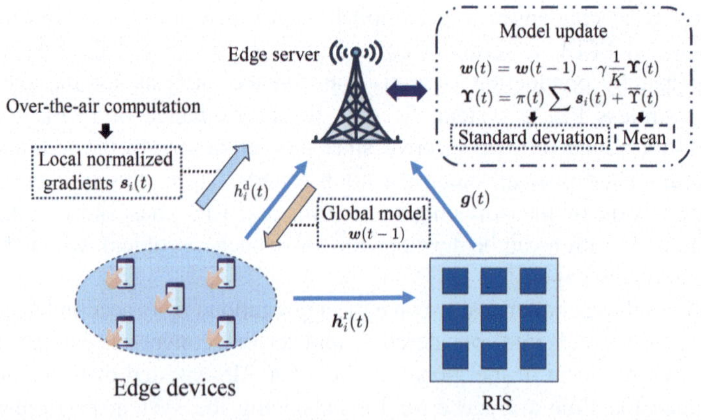

Fig. 5.1 Illustration of model aggregation in an RIS-assisted over-the-air FEEL system

than that of edge devices, we assume that channel fading and noise during the downlink transmission can be fully compensate.

- *Local model update*: After receiving global model $\boldsymbol{w}(t-1)$, each edge device $k \in \mathcal{K}$ computes its local stochastic gradient

$$\boldsymbol{\Upsilon}_k(t) = \nabla F_k\left(\boldsymbol{w}(t-1); \mathcal{B}_k^{FEEL}\right),$$

where $\mathcal{B}_k^{FEEL} \subset \mathcal{D}_k$ denotes the mini-batch containing $|\mathcal{B}_k^{FEEL}|$ randomly selected data samples.

- *Local model aggregation*: Since only the average of local gradients (i.e., $\boldsymbol{\Upsilon}(t) = \frac{1}{K}\sum_{k\in\mathcal{K}} \boldsymbol{\Upsilon}_k(t)$) is required at the edge server for the global model update (i.e., $\boldsymbol{w}(t)$), we adopt AirComp for the spectrum-efficient gradient aggregation. With AirComp, the concurrently transmitted local gradients $\{\boldsymbol{\Upsilon}_k(t)\}_{k\in\mathcal{K}}$ from K edge devices can be added over the air, and the edge server can directly receive a summation of these gradients. With the existence of random channel fading and receiver noise, the edge server recovers a noisy estimation of the average gradient $\boldsymbol{\Upsilon}(t)$. The achievable accuracy of local model aggregation via AirComp, however, is limited by the worst channel among all device-server links [6]. To address this problem, an RIS with N reflection elements is leveraged to mitigate the communication bottleneck and in turn improve the gradient aggregation accuracy. Denote the recovered noisy estimation of $\boldsymbol{\Upsilon}(t)$ as $\hat{\boldsymbol{\Upsilon}}(t)$, the edge server update the global model as

$$\boldsymbol{w}(t) = \boldsymbol{w}(t-1) - \gamma\hat{\boldsymbol{\Upsilon}}(t), \tag{5.1}$$

where γ denotes the learning rate.

5.2.2 Signal Model

After the local training, each edge device $k \in \mathcal{K}$ calculates the mean $\bar{\Upsilon}_k(t)$ and variance $\pi_k^2(t)$ of local gradient $\boldsymbol{\Upsilon}_k(t)$ as $\bar{\Upsilon}_k(t) = \frac{1}{\Omega}\sum_{j=1}^{\Omega} \Upsilon_{k,j}(t)$ and $\pi_k^2(t) = \frac{1}{\Omega}\sum_{j=1}^{\Omega}\left(\Upsilon_{k,j}(t) - \bar{\Upsilon}_k(t)\right)^2$, respectively, where $\Upsilon_{k,j}(t)$ is the j-th element of $\boldsymbol{\Upsilon}_k(t)$. After receiving the local statistics $\{\bar{\Upsilon}_k(t), \pi_k^2(t)\}$, the edge server computes the global mean and variance as $\bar{\Upsilon}(t) = \frac{1}{K}\sum_{j=1}^{K} \bar{\Upsilon}_k(t)$ and $\pi^2(t) = \frac{1}{K}\sum_{j=1}^{K} \pi_k^2(t)$, respectively, and then broadcasts them to each edge device k for the normalization of $\boldsymbol{\Upsilon}_k(t)$ as in [12]. Each edge device normalizes its transit signal as $\boldsymbol{s}_k(t) = \frac{\boldsymbol{\Upsilon}_k(t) - \bar{\Upsilon}_k(t)}{\pi(t)}$ to satisfy $\mathbb{E}[\boldsymbol{s}_k(t)] = 0$ and $\mathbb{E}[\boldsymbol{s}_k(t)\boldsymbol{s}_k(t)^{\mathrm{T}}] = \boldsymbol{I}_{\Omega}$. Generally, the gradient statistics $\{\bar{\Upsilon}_k(t), \pi_k^2(t)\}$ have a much smaller size than the gradient vector [12]. Thus, the exchange of $\{\bar{\Upsilon}_k(t), \pi_k^2(t)\}$ is assumed to be error-free.

We denote $\boldsymbol{g}(t) \in \mathbb{C}^N$, $\boldsymbol{h}_i^{\mathrm{r}}(t) \in \mathbb{C}^N$, and $h_i^{\mathrm{d}}(t) \in \mathbb{C}$ as the channel between the RIS and edge server, channel between device i and RIS, and channel between device i and edge server, respectively. We assume the above channels to obey block-fading [1]. We denote $\boldsymbol{v}(t) = [\beta e^{j\theta_1(t)}, \beta e^{j\theta_2(t)}, \ldots, \beta e^{j\theta_N(t)}]^{\mathrm{T}}$ as the phase-shift vector of RIS with β being the reflection amplitude and $\theta_n(t)$ being the reflection phase of the n-th element. For simplicity, we set β to 1, as in [12, 18, 21]. With synchronized transmissions, the signal received at the edge server in round t is

$$\boldsymbol{y}(t) = \sum_{i=1}^{K} \left[h_i^{\mathrm{d}}(t) + \boldsymbol{g}(t)\mathrm{diag}\left(\boldsymbol{v}(t)\right)\boldsymbol{h}_i^{\mathrm{r}}(t)\right] b_i(t)\boldsymbol{s}_i(t) + \boldsymbol{n}(t), \tag{5.2}$$

where $b_i(t)$ denotes the transmit scalar of device i, and $\boldsymbol{n}(t) \sim \mathcal{CN}(0, \sigma^2\boldsymbol{I})$ denotes the *additive white Gaussian noise* (AWGN).

For the simplicity of notation, we define $h_i^{\mathrm{c}}(t) = h_i^{\mathrm{d}}(t) + \boldsymbol{g}(t)\mathrm{diag}(\boldsymbol{v}(t))\boldsymbol{h}_i^{\mathrm{r}}(t)$ as the combined channel response between device i and edge server. Moreover, we use $p_i(t) \geq 0$ to denote the transmit power of device i. Given that various channel estimation strategies have been proposed in the literature for RIS-assisted wireless networks, such as those in [13, 14, 17, 24], we assume that perfect CSI is available in this paper. Besides, we ignore the communication overhead resulting from channel estimation, since the pilot length is usually much smaller than the size of the gradient vector. By setting $b_i(t) = \frac{\sqrt{p_i(t)}(h_i^{\mathrm{c}}(t))^{\dagger}}{|h_i^{\mathrm{c}}(t)|}$, the received signal in (5.2) is $\boldsymbol{y}(t) = \sum_{i=1}^{K}\left(|h_i^{\mathrm{c}}(t)|\sqrt{p_i(t)}\boldsymbol{s}_i(t)\right) + \boldsymbol{n}(t)$. Besides, we consider a threshold $\bar{P}_k$ on the average transmit power over T rounds at device k, i.e., $\frac{1}{T}\sum_{t=1}^{T} p_k(t) \leq \bar{P}_k, \forall\, k \in \mathcal{K}$. After receiving signal $\boldsymbol{y}(t)$, the edge server recovers the arithmetic mean of $\boldsymbol{s}_i(t)$ by adopting the denoising factor $\eta(t) > 0$ as follows

$$\hat{\boldsymbol{s}}(t) = \frac{\boldsymbol{y}(t)}{\sqrt{\eta(t)}} = \sum_{i\in\mathcal{K}} \frac{\sqrt{p_i(t)}|h_i^{\mathrm{c}}(t)|}{\sqrt{\eta(t)}}\boldsymbol{s}_i(t) + \frac{\boldsymbol{n}(t)}{\sqrt{\eta(t)}}. \tag{5.3}$$

As $\boldsymbol{\Upsilon}_k(t) = \pi(t)\boldsymbol{s}_k(t) + \bar{\Upsilon}(t)$ and $\boldsymbol{s}(t) = \sum_{k\in\mathcal{K}}\boldsymbol{s}_k(t)$, the received stochastic gradient is $\hat{\boldsymbol{\Upsilon}}(t) = \frac{1}{K}\pi(t)\left(\hat{\boldsymbol{s}}(t) - \boldsymbol{s}(t)\right) + \boldsymbol{\Upsilon}(t)$, and the induced error in the uplink is

$$\bar{\boldsymbol{e}}(t) = \frac{1}{K}\left(\hat{\boldsymbol{s}}(t) - \boldsymbol{s}(t)\right). \tag{5.4}$$

By substituting (5.3) into (5.4), we have

$$\bar{\boldsymbol{e}}(t) = \frac{1}{K}\sum_{i\in\mathcal{K}}\left(\frac{\sqrt{p_i(t)}|h_i^{\mathrm{c}}(t)|}{\sqrt{\eta(t)}} - 1\right)\boldsymbol{s}_i(t) + \frac{1}{K}\frac{\boldsymbol{n}(t)}{\sqrt{\eta(t)}}. \tag{5.5}$$

It is worth noting that FEEL performance heavily relies on a range of factors, including the transmit power, channel conditions, RIS phase shifts, denoising factor, and receiver noise. The inherent influence of these underlying parameters on the accuracy of local gradient aggregation can significantly impact the overall FEEL performance. In the subsequent analysis, we delve deeper into how each of these factors affects the efficacy of FEEL in practical scenarios.

5.3 Convergence Analysis and Problem Formulation

We analyze the convergence behavior of the over-ther-air assisted FEEL in this section.

5.3.1 Basic Assumptions

To make the analysis tractable, we make the following assumptions:

Assumption 8 *The global loss function, i.e.,* $F(\boldsymbol{w}) \geq F(\mathbf{w}^*)$, $\forall$ $\mathbf{w}$ *is lower bounded by* $F(\mathbf{w}^*)$.

Assumption 9 $F_k(\boldsymbol{w})$ *is S-smooth*

$$\|\nabla F_k(\boldsymbol{w}) - \nabla F_k(\mathbf{w}')\|_2 \leq S\|\mathbf{w} - \mathbf{w}'\|_2, \ \forall \ \mathbf{w}, \mathbf{w}'. \tag{5.6}$$

Assumption 10 *The local mini-batch gradient* $\boldsymbol{\Upsilon}_k$ *is an unbiased estimate of* $\nabla F_k(\boldsymbol{w})$, *and its variance is bounded with a constant* $\xi \geq 0$, *i.e.,* $\mathbb{E}[\boldsymbol{\Upsilon}_k] = \nabla F_k(\boldsymbol{w})$ *and* $Var(\boldsymbol{\Upsilon}_k) = \mathbb{E}[\|\boldsymbol{\Upsilon}_k - \nabla F_k(\boldsymbol{w})\|_2^2] \leq \xi^2$.

Assumption 11 *There is a constant* $\Gamma \geq 0$ *that upper bounds the variance of* Ω *elements of* $\boldsymbol{\Upsilon}_k$.

5.3.2 Convergence Results

With the above assumptions, we obtain Lemma 5.1.

Lemma 5.1 *The expected norm of the induced error is upper bounded as*

$$\mathbb{E}[\|\boldsymbol{e}(t)\|_2^2] \leq \Omega \frac{\Gamma(K+1)}{K^2} \mathrm{Er}(t), \tag{5.7}$$

where

$$\mathrm{Er}(t) = \sum_{k=1}^{K} \left(\frac{\sqrt{p_k(t)}|h_i^{\mathrm{d}}(t) + \boldsymbol{g}(t)\mathrm{diag}(\boldsymbol{v}(t))\boldsymbol{h}_i^{\mathrm{r}}(t)|}{\sqrt{\eta(t)}} - 1 \right)^2 + \frac{\sigma^2}{\eta(t)}.$$

According to Lemma 5.1, we derive an upper bound of $\frac{1}{T}\sum_{t=0}^{T-1} \|\nabla F(\boldsymbol{w}(t))\|_2^2$ in Theorem 5.1.

Theorem 5.1 (Convergence) *With Assumptions 1–4, if* $\gamma < \frac{1}{2S}$ *and after* T *rounds, we bound the time-average norm of the global gradient as*

$$\frac{1}{T}\sum_{t=0}^{T-1} \|\nabla F(\boldsymbol{w}(t))\|_2^2 \leq \underbrace{\frac{2(F(\mathbf{w}(0) - F(\mathbf{w}^*)))}{\gamma(1-2\gamma S)T}}_{\textit{Initial optimality gap}} + \underbrace{\frac{2S\gamma\xi^2}{K(1-2\gamma S)}}_{\textit{Gradient variance induced gap}}$$

$$+ \underbrace{\frac{1+2\gamma S}{1-2\gamma S}\frac{1}{T}\sum_{t=0}^{T-1} \underbrace{\mathbb{E}\left[\|\bar{\boldsymbol{e}}(t)\|_2^2\right]}_{\textit{Instantaneous error}}}_{\textit{Time-average error}}.$$

5.3.3 *Upper Bound Decomposition*

Theorem 5.1 establishes an upper bound that comprises the initial optimality gap, gradient variance induced gap, and time-average error. As $T \to \infty$, the initial optimality gap tends to zero. The gradient variance induced gap is influenced by various factors like the learning rate, Lipschitz constant, number of edge devices, and local gradient variance. Increasing the mini-batch size helps reduce the gradient variance induced gap, since it can be bounded with a smaller value of ξ. Moreover, involving more edge devices also decreases the gap. When γ, ξ, and S are fixed, the gradient variance induced gap remains a constant, and the time-average error becomes dominant in the upper bound with a large T. Therefore, there is a need to minimize the time-average error to enhance FEEL convergence performance.

The convergence of FEEL is impaired by transmission errors. Therefore, it is crucial to formulate a joint optimization problem that minimizes the time-average error while considering practical constraints such as hardware restrictions, communication limitations, and power constraints. This ensures that the optimized solution can be implemented effectively in a real-world FEEL system.

5.3.4 Problem Formulation

Our objective is to present an algorithm that can efficiently scale and reduce the time-average error. The corresponding optimization problem is

$$\begin{aligned}\mathscr{P}_0: \min_{\substack{\{\boldsymbol{p}(t)\}\\ \{\boldsymbol{v}(t),\eta(t)\}}} \quad & \frac{1}{T}\sum_{t=1}^{T}\left[\sum_{k\in\mathcal{K}}\left(\frac{\sqrt{p_k(t)}|h_k^{\mathrm{c}}(t)|}{\sqrt{\eta(t)}}-1\right)^2+\frac{\sigma^2}{\eta(t)}\right]\\ \text{s.t.} \quad & \frac{1}{T}\sum_{t=1}^{T} p_k(t)\le \bar{P}_k, \forall\, k\in\mathcal{K},\\ & |v_j(t)|=1, \forall\, j=1,\dots,N,\ \ \forall\, t=1,\dots,T,\end{aligned} \tag{5.8}$$

where $\boldsymbol{p}(t) = [p_1(t), \dots, p_K(t)]$ denotes the transmit power vector of edge devices.

To solve problem $\mathscr{P}_0$, we first develop an alternating optimization algorithm in Sect. 5.4, and then propose a novel GNN-based learning algorithm in Sect. 5.5.

5.4 Alternating Optimization Algorithm

5.4.1 AirComp Transceiver Optimization

Problem $\mathscr{P}_0$ meets the time-sharing condition [23] and strong duality, and thus the Lagrangian-duality method can be applied. The Lagrangian function of problem $\mathscr{P}_0$ is

$$\mathcal{L}(\{\boldsymbol{p}(t), \boldsymbol{v}(t), \eta(t), \lambda_k\}) = \frac{1}{T}\sum_{t=1}^{T}\mathrm{Er}(t) + \sum_{k\in\mathcal{K}}\lambda_k\left(\frac{1}{T}\sum_{t=1}^{T}p_k(t) - \bar{P}_k\right),$$

where $\lambda_k \geq 0$ represents the dual variable with respect to the transmit power constraint of device k. The Lagrangian dual function and the Lagrangian dual problem can be respectively expressed as

$$G(\{\lambda_k\}) = \min_{\{\boldsymbol{p}(t),\boldsymbol{v}(t),\eta(t)\}} \mathcal{L}(\{\boldsymbol{p}(t), \boldsymbol{v}(t), \eta(t), \lambda_k\}), \tag{5.9}$$

$$\max_{\{\lambda_k\geq 0\}} G(\{\lambda_k\}), \ \forall\, k \in \mathcal{K}. \tag{5.10}$$

As the strong duality holds, problem $\mathscr{P}_0$ can be equivalently solved by maximizing $G(\{\lambda_k\})$, which is derived by solving the minimization problem in (5.9) for given $\{\lambda_k \mid k \in \mathcal{K}\}$. We decompose problem (5.9) into multiple sub-problems as follows

$$\min_{\substack{\{\boldsymbol{p}(t),\boldsymbol{v}(t),\}\\ \eta(t),\lambda_k}} \sum_{k\in\mathcal{K}} \left(\frac{\sqrt{p_k(t)}|h_k^{\mathrm{c}}(t)|}{\sqrt{\eta(t)}} - 1 \right)^2 + \frac{\sigma^2}{\eta(t)} + \sum_{k\in\mathcal{K}} \lambda_k p_k(t). \tag{5.11}$$

Referring to [1], the transmit power and the denoising factor with the given phase-shift vector $\boldsymbol{v}(t)$ can be set as

$$p_k^*(t) = \left(\frac{\sqrt{|h_k^{\mathrm{c}}(t)|^2 \eta(t)}}{|h_k^{\mathrm{c}}(t)|^2 + \eta(t)\lambda_k} \right)^2, \tag{5.12}$$

where $\eta(t)$ can be optimized by applying bisection search to solve

$$\sum_{k\in\mathcal{K}} \frac{\zeta_k(t)}{(\zeta_k(t)\nu(t)+1)^2} = \sigma^2 \tag{5.13}$$

with $\zeta_k(t) = \frac{|h_k^{\mathrm{c}}(t)|^2}{\lambda_k}$ and $\nu(t) = \frac{1}{\eta(t)}$. The dual variables $\{\lambda_k\}_{k\in\mathcal{K}}$ can be optimized via a sub-gradient based method. Specifically, to maximize $G(\{\lambda_k\})$, we can update the dual variables $\{\lambda_k\}_{k\in\mathcal{K}}$ as follows

$$\lambda_k \leftarrow \lambda_k + m \left(\frac{1}{T} \sum_{t=1}^{T} p_k(t) - \bar{P}_k \right), \ \forall\, k \in \mathcal{K}, \tag{5.14}$$

where $\frac{1}{T}\sum_{t=1}^{T} p_k(t) - \bar{P}_k$ is the sub-gradient of $G(\{\lambda_k\})$, $\forall\, k \in \mathcal{K}$ and m is the step size.

5.4.2 RIS Phase-Shift Optimization

When $\boldsymbol{p}(t)$ and $\eta(t)$ fixed, problem (5.11) can be rewritten as

$$\begin{aligned} \mathscr{P}_1 :\ & \min_{\boldsymbol{v}(t)} : \phi(\boldsymbol{v}(t)) \\ & \text{s.t. } |\boldsymbol{v}_j(t)| = 1, \forall\, j = 1, \ldots, N, \ \ \forall\, t = 1, \ldots, T, \end{aligned}$$

where

$$\phi(\boldsymbol{v}(t)) = \sum_{k=1}^{K} \left[\frac{p_k(t)|h_k^{\mathrm{d}}(t) + \boldsymbol{a}_k(t)\boldsymbol{v}(t)|^2}{\eta} - 2\sqrt{\frac{p_k(t)}{\eta(t)}}|h_k^{\mathrm{d}}(t) + \boldsymbol{a}_k(t)\boldsymbol{v}(t)| \right], \tag{5.15}$$

the constant term $\sum_{k\in\mathcal{K}} \lambda_k p_k(t)$ is ignored, and $\boldsymbol{a}_k(t) = \boldsymbol{g}(t)\mathrm{diag}(\boldsymbol{h}_k^{\mathrm{r}}(t))$. Defining $\breve{\boldsymbol{v}}(t) = [\boldsymbol{v}(t), 1]^T$, $\boldsymbol{r}_k(t) = [\boldsymbol{a}_k(t), h_k^{\mathrm{d}}(t)]$, and matrices $\boldsymbol{R}_k(t) = \boldsymbol{r}_k^{\mathrm{H}}(t)\boldsymbol{r}_k(t)$, we can reformulate problem $\mathscr{P}_1$ as

$$\begin{aligned} \mathscr{P}_2 : \min_{\breve{\boldsymbol{v}}(t)} & \sum_{k=1}^{K} \left(\sqrt{\frac{p_k(t)}{\eta(t)} \breve{\boldsymbol{v}}^{\mathrm{H}}(t)\boldsymbol{R}_k(t)\breve{\boldsymbol{v}}(t)} - 1 \right)^2 \\ \text{s.t. } & |\boldsymbol{v}_j(t)| = 1, \forall\, j = 1, \ldots, N, \end{aligned} \tag{5.16}$$

which is non-convex due to the square root operation. Considering the fact that the optimal $\breve{\boldsymbol{v}}$ for problem $\mathscr{P}_2$ forces $\sqrt{\frac{p_k(t)}{\eta(t)} \breve{\boldsymbol{v}}^{\mathrm{H}}(t)\boldsymbol{R}_k(t)\breve{\boldsymbol{v}}(t)}$ to approach 1, we use a convex function with respect to $\breve{\boldsymbol{v}}$ as $\frac{p_k(t)}{\eta(t)} \breve{\boldsymbol{v}}^{\mathrm{H}}(t)\boldsymbol{R}_k(t)\breve{\boldsymbol{v}}(t)$ to approximate the square root term $\sqrt{\frac{p_k(t)}{\eta(t)} \breve{\boldsymbol{v}}^{\mathrm{H}}(t)\boldsymbol{R}_k(t)\breve{\boldsymbol{v}}(t)}$. Then, problem $\mathscr{P}_2$ can be converted to the following non-convex quadratically constrained quadratic programming (QCQP) problem

$$\begin{aligned} \mathscr{P}_3 : \min_{\breve{\boldsymbol{v}}(t)} & \sum_{k=1}^{K} \left(\frac{p_k}{\eta} \breve{\boldsymbol{v}}^{\mathrm{H}}\boldsymbol{R}_k\breve{\boldsymbol{v}} - 1 \right)^2 \\ \text{s.t. } & |\breve{\boldsymbol{v}}_j(t)|^2 = 1, \quad \forall\, j = 1, \ldots, N, \quad \forall\, t = 1, \ldots, T, \end{aligned}$$

where $\breve{\boldsymbol{v}}_n(t)$ is the n-th element of $\breve{\boldsymbol{v}}(t)$. To convexify problem $\mathscr{P}_3$, we adopt matrix lifting to linearize problem $\mathscr{P}_3$. By denoting $\breve{\boldsymbol{v}}^{\mathrm{H}}(t)\boldsymbol{R}_k(t)\breve{\boldsymbol{v}}(t)$ as $\mathrm{Tr}(\boldsymbol{R}_k(t)\boldsymbol{V}(t))$ with $\boldsymbol{V}(t) = \breve{\boldsymbol{v}}(t)\breve{\boldsymbol{v}}^{\mathrm{H}}(t)$, problem $\mathscr{P}_3$ can be rewritten as

$$\begin{aligned} \mathscr{P}_4 : \min_{\boldsymbol{V}} & \sum_{k=1}^{K} \left(\frac{p_k(t)}{\eta(t)} \mathrm{Tr}(\boldsymbol{R}_k(t)\boldsymbol{V}(t)) - 1 \right)^2 \\ \text{s.t. } & \boldsymbol{V}_{i,i}(t) = 1, \quad \forall i \, = 1, \ldots, N+1, \\ & \mathrm{rank}(\boldsymbol{V}(t)) = 1, \\ & \boldsymbol{V}(t) \succeq 0, \quad \forall\, t = 1, \ldots, T. \end{aligned} \tag{5.17}$$

For the positive semi-definite (PSD) matrix $\boldsymbol{V}$, constraint $\mathrm{rank}(\boldsymbol{V}(t)) = 1$ is equivalent to $\mathrm{Tr}(\boldsymbol{V}(t)) - \|\boldsymbol{V}(t)\|_2 = 0$, where $\|\boldsymbol{V}(t)\|_2$ denotes the spectral norm of matrix $\boldsymbol{V}(t)$. Hence, problem $\mathscr{P}_4$ can be reformulated as

$$\begin{aligned}\mathscr{P}_5 : \min_{\boldsymbol{V}(t)} & \sum_{k=1}^{K} \left(\frac{p_k(t)}{\eta(t)} \mathrm{Tr}\left(\boldsymbol{R}_k(t)\boldsymbol{V}(t)\right) - 1 \right)^2 + \mu(t) \\ \text{s.t. } & \boldsymbol{V}_{i,i}(t) = 1,\ \forall i = 1, \ldots, N+1, \\ & \boldsymbol{V}(t) \succeq 0,\ \forall t = 1, \ldots, T,\end{aligned} \tag{5.18}$$

where $\mu(t) = \rho(\mathrm{Tr}(\boldsymbol{V}(t)) - \|\boldsymbol{V}(t)\|_2)$.

Problem $\mathscr{P}_5$ is a semi-definite programming (SDP) problem. We can obtain an exact rank-one optimal solution $\boldsymbol{V}^*(t)$ via forcing the difference between trace norm and spectral norm to be zero. Subsequently, we can obtain a feasible solution $\check{\boldsymbol{v}}(t)$ via the Cholesky decomposition, i.e., $\boldsymbol{V}^*(t) = \check{\boldsymbol{v}}^{\mathrm{H}}(t)\check{\boldsymbol{v}}(t)$. On the other hand, problem $\mathscr{P}_5$ is considered to be infeasible if the objective value of (5.18) fails to be zero.

By now, we introduce an alternating optimization method to solve problem $\mathscr{P}_0$. In each iteration, we update the phase-shift vector by solving the convex problem $\mathscr{P}_5$. After updating the combined channel coefficients with the optimized $\boldsymbol{v}^*(t)$, we can update $\boldsymbol{p}^*(t)$, $\eta^*(t)$, and $\{\lambda_k^*\}_{k\in\mathcal{K}}$ in an alternating manner by utilizing (5.12), (5.13), and (5.14), respectively.

5.4.3 Computation Complexity

We tackle problem $\mathscr{P}_0$ by solving a series of SDP problems. Solving each SDP problem is of the computational complexity $\mathcal{O}((N+1)^{4.5}\log(1/\epsilon_2^{\mathrm{bis}}))$, where $\epsilon_2^{\mathrm{bis}}$ denotes the accuracy. Therefore, by denoting the umber of the SDR problems as J, the computational complexity for RIS phase-shifts optimization is expressed as $\mathcal{O}(J(N+1)^{4.5}\log(1/\epsilon_2^{\mathrm{bis}}))$. For the optimization of transmit power and denoising factor, the main iteration lies in the bisection search for optimal η, thereby the computational complexity is $\mathcal{O}(\log(1/\epsilon_1^{\mathrm{bis}}))$, where $\epsilon_1^{\mathrm{bis}}$ is the accuracy of bisection search. Hence, the overall computational complexity of the alternating optimization method is $\mathcal{O}(G(J(N+1)^{4.5}\log(1/\epsilon_2^{\mathrm{bis}}) + \log(1/\epsilon_1^{\mathrm{bis}})))$, where G is the number of alternating iterations. The computational complexity is dominated by the optimization of RIS, the complexity of which grows exponentially with the number of RIS elements.

5.5 GNN-Based Learning Algorithm

In this section, we develop a novel GNN-based learning algorithm to solve problem $\mathscr{P}_0$, and elaborate on the corresponding architecture design and training of neural networks.

5.5.1 Graphical Representation

In order to overcome the restrictions posed by the optimization-based algorithm, we have devised a novel learning framework that uses GNN technology. This framework enables us to directly map the channel coefficients to the optimal parameter setting, resulting in more efficient and accurate results. Denoting the mapping function by $\kappa(\cdot)$, which maps the channel coefficients (i.e., $\{h_i^{\mathrm{d}}(t)\}$ and $\{\boldsymbol{g}(t)\mathrm{diag}(\boldsymbol{h}_i^{\mathrm{r}}(t))\}$) to devices' transmit powers (i.e., $\boldsymbol{p}(t)$), denoising factor (i.e., $\eta(t)$), and RIS phase-shift vector (i.e., $\boldsymbol{v}(t)$). Hence, the mapping function can be expressed as

$$\{\boldsymbol{p}(t), \boldsymbol{v}(t), \eta(t)\} = \kappa(h_i^{\mathrm{d}}(t),\ \boldsymbol{g}(t)\mathrm{diag}(\boldsymbol{h}_i^{\mathrm{r}}(t))), \tag{5.19}$$

and solving problem $\mathscr{P}_0$ can be interpreted as learning the optimal mapping function $\kappa(\cdot)$. By utilizing the universal approximation characteristic inherent in deep neural networks (DNN), we can effectively parameterize the mapping function $\kappa(\cdot)$, which is responsible for linking the channel coefficients to the optimal parameter setting. We then adopt a data-driven approach to train the neural networks and develop a sophisticated mapping function that produces optimum results.

Given that problem $\mathscr{P}_0$ necessitates the simultaneous minimization of signal misalignment error and receiver noise, a coordinated approach must be taken to optimize the AirComp transceiver design (i.e., transmit power and denoising factor) as well as the RIS phase-shifts. To achieve this objective, we have developed a GNN-based learning framework that is capable of enhancing learning performance by training the mapping function for optimal results. Our graph comprises $K+2$ nodes and $2K+1$ edges that connects relevant nodes. Specifically, nodes 1 to K represent K edge devices, while node $K+1$ and node $K+2$ represent the edge server and RIS, respectively. Each node is assigned with a representation vector denoted as z_k, which is trained to contain all the relevant information required for establishing an effective mapping. We update representation vectors in a layer-wise manner, utilizing combining and aggregation operations that take the preceding layer's representation vectors as input. These operations will be explained more thoroughly in the next subsection. After updating, all representation vectors should contain enough information for the joint design of AirComp transceiver and RIS phase-shifts. Specifically, the optimized setting of the AirComp transceiver can be directly obtained from representation vectors $z_1, \ldots, z_{K+1}$, while that of the RIS can be obtained from representation vector z_{K+2}.

5.5.2 GNN Architecture Design

This section provides a detailed overview of the GNN-based learning algorithm proposed in this work. The algorithm is designed to be executed at the edge server, which receives the channel coefficients and uses them to derive an optimized AirComp transceiver and RIS phase-shifts design. The optimized design parameters are then fed back to the edge devices and RIS controller for uplink model aggregation. The GNN consists of an initialization layer, several graphical mapping layers, and one parameter generation layer. In the initialization layer, the channel coefficients are transformed into representation vectors $z_k^{(0)}, \forall\, k = 1, \ldots, K+2$. Subsequently, all nodes' representation vectors are updated via D graphical mapping layers, resulting in $z_k^{(D)}, \forall\, k = 1, \ldots, K+2$. Finally, the parameter generation layer is used to obtain the learned parameters of the AirComp transceiver (i.e., $\{\eta(t), \boldsymbol{p}(t)\}$) and the RIS phase-shifts (i.e., $\boldsymbol{v}(t)$). Figure 5.2 provides an illustration of the overall network architecture. Each of these layers' design and functionality will be explicitly explained below.

- *Initialization layer:* To derive a set of initial representation vectors $z_k^{(0)}$ required for further processing, the initialization layer initiates the channel coefficients into representation vectors. An encoder function $f_{\mathrm{EC}}^0(\cdot)$ is employed in this layer, which comprises three linear layers of a multi-layer perceptron (MLP). To make this MLP effective, activation functions and batch normalization are introduced between adjacent linear layers. The rectified linear unit (ReLU) activation function with the expression $\mathrm{ReLU}(x) = \max(0, x)$ is chosen because it provides a linear mapping property during back-propagation calculations and has a low complexity. Since deep learning toolkits do not fully support complex values in channel coefficients, we divide them into real and imaginary components before feeding them to the encoder function. As a result of this division, we can accurately represent each edge device node $k \in \mathcal{K}$ by obtaining a precise representation vector $z_k^{(0)}$ based on its corresponding channel coefficients.

$$z_k^{(0)} = f_{\mathrm{EC}}^0\left(\left[\Re(h_k^{\mathrm{d}}(t) + \boldsymbol{g}(t)\boldsymbol{h}_k^{\mathrm{r}}(t)), \Im(h_k^{\mathrm{d}}(t) + \boldsymbol{g}(t)\boldsymbol{h}_k^{\mathrm{r}}(t))\right]\right).$$

  Permutation equivariance can be ensured by utilizing a shared encoder among edge device nodes, which is an important aspect to consider. This results in the reduction of traninable parameters and also decreases the computational complexity during forward inference. We determine the representation vectors of the RIS and edge server, denoted as $f_{K+2}^0(\cdot)$ and $f_{K+1}^0(\cdot)$ respectively, as

$$z_{K+1}^{(0)} = f_{K+1}^0(\bar{z}^{(0)}), \quad z_{K+2}^{(0)} = f_{K+2}^0(\bar{z}^{(0)}), \tag{5.20}$$

  where $\bar{z}^{(0)} = \frac{1}{K}\sum_{k=1}^{K} z_k^{(0)}$ is the average representation vector of all the edge device nodes. The average representation is utilized because the final representation vectors for the edge server and RIS are a collective contribution from the representation vectors of all device nodes.
- *Graphical mapping layer:* In the graphical mapping model, each layer comprises $K+2$ modules catering to $K+2$ nodes. These modules work to gather and amalgamate data from the representation vectors of all nodes presented in the

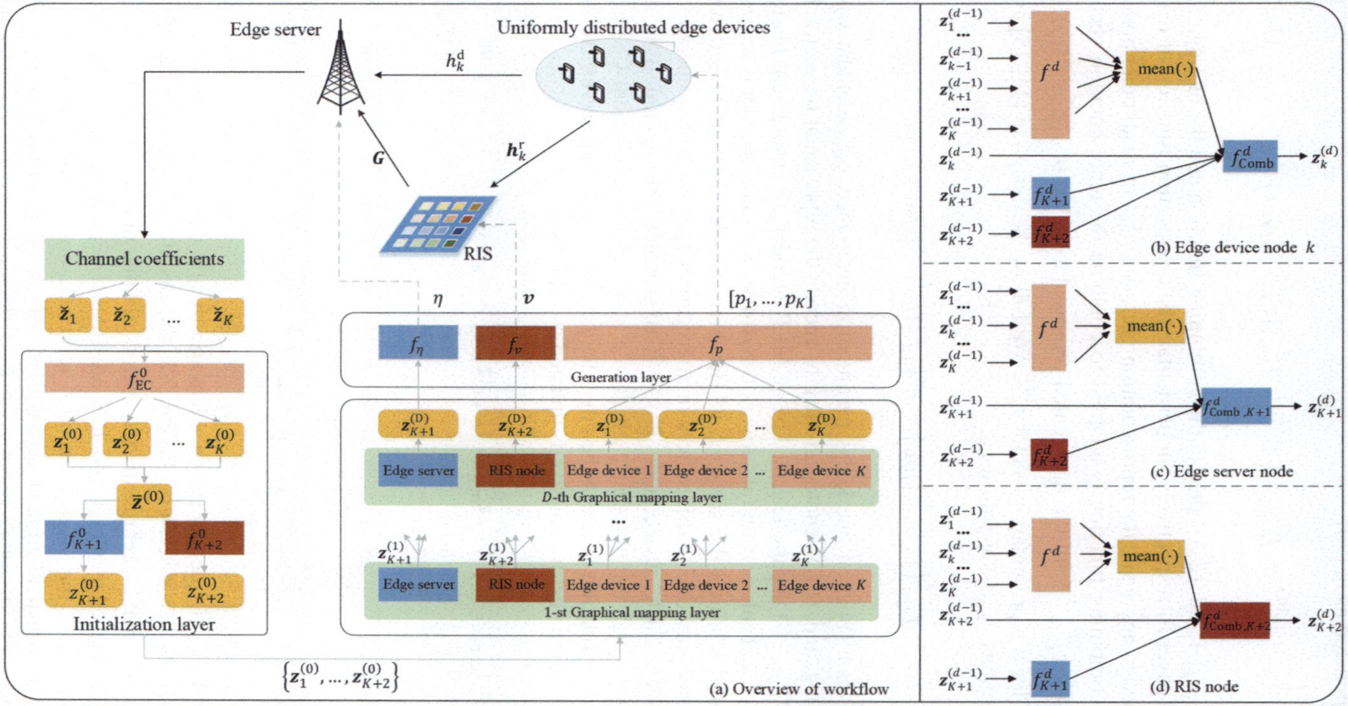

Fig. 5.2 (**a**) Overall workflow of the proposed GNN-based learning algorithm for RIS-assisted over-the-air FEEL; (**b**) Module design of the d-th graphical mapping layer for edge device nodes; (**c**) Module design of the d-th graphical mapping layer for edge server node; (**d**) Module design of the d-th graphical mapping layer for RIS

previous layer. In general, the update of each node in the d-th graphical mapping layer is conducted as in [20]

$$z_k^{(d)} = f_{\mathrm{comb},k}^d \left(z_k^{(d-1)}, f_{\mathrm{agg}}^d(\{z_j^{(d-1)}\}_{j \neq k}) \right), \tag{5.21}$$

where $f_{\mathrm{comb},k}^d(\cdot)$ and $f_{\mathrm{agg}}^d(\cdot)$ denote the combining function and the aggregation function at layer d, respectively. In particular, the aggregation function of node k is designed to aggregate the representation vectors $z_j^{(d-1)}$ from all other nodes, while the combining function is adopted to combine the representation vector of local node $z_k^{(d-1)}$ with the aggregated representation vectors $\{z_j^{(d-1)}\}_{j \neq k}$ to update the local representation vector $z_k^{(d)}$. We follow the design in [11, 15] and design functions $f_{\mathrm{agg}}(\cdot)$ as follows

$$f_{\mathrm{agg}}^d \left(\{z_j^{(d-1)}\}_{j \neq k} \right) = \phi \left(\{f^d(z_j^{(d-1)})\}_{j \neq k} \right),$$

where $\phi(\cdot)$ is a function that keeps permutation invariance, e.g., element-wise mean pooling, and $f^d(\cdot)$ denotes an MLP to encode the representation vector. In addition, $f_{\mathrm{comb},k}^d(\cdot)$ is parameterized by an MLP encoder. To ensure permutation equivariance and enhance neural network robustness, the same encoder is utilized for all modules within the same layer. However, detailed designs of different modules may vary due to differences in aggregation and combining operations among nodes. Each device node $k \in \mathcal{K}$ aggregates information from all other nodes, which ensures that each device node acquires sufficient CSI and facilitates coordination between the edge devices, RIS, and edge server. Hence, the representation vector at node k is updated as follows

$$z_k^{(d)} = f_{\mathrm{comb},k}^d \left(z_k^{(d-1)}, z_{\mathrm{agg},k}^{(d)}, f_{K+1}^d(z_{K+1}^{(d-1)}), f_{K+2}^d(z_{K+2}^{(d-1)}) \right), \forall k \in \mathcal{K}, \tag{5.22}$$

where

$$z_{\mathrm{agg},k}^{(d)} = \frac{1}{K-1} \sum_{j \neq k}^{1 \leq j \leq K} \left(f^d(z_j^{(d-1)}) \right), \tag{5.23}$$

$f_{K+1}^d(\cdot)$ and $f_{K+2}^d(\cdot)$ denote the encoders specifically designed for node $K+1$ and node $K+2$, respectively. In Fig. 5.2, the update process is depicted. It should be noted that the representation vectors of other device nodes, i.e., $z_j^{(d-1)}$, and those of the edge server and RIS are treated separately in (5.22). Node $k \in \mathcal{K}$ aggregates the average encoded representation vector of other device nodes as the signal alignment depends on all channels instead of only the strongest one. On the other hand, node $k \in \mathcal{K}$ aggregates the representation vectors of the RIS node and edge server node, which do not alter their permutation invariance property. This design enables the neural network to learn the channel representation better with respect to its channel information at both the RIS node and the edge server node. Meanwhile, the update of the representation vectors of the RIS node and the edge server node can be expressed as $z_{K+1}^{(d)} = f_{\text{Comb},K+1}^{d}\left(z_{\text{agg}}^{(d)}, f_{K+2}^{d}(z_{K+2}^{(d-1)}), z_{K+1}^{(d-1)}\right)$ and $z_{K+2}^{(d)} = f_{\text{Comb},K+2}^{d}\left(z_{\text{agg}}^{(d)}, f_{K+1}^{d}(z_{K+1}^{(d-1)}), z_{K+2}^{(d-1)}\right)$, respectively, where

$$z_{\text{agg}}^{(d)} = \frac{1}{K}\sum_{i=1}^{K} f^{d}(z_i^{(d-1)}). \tag{5.24}$$

Figure 5.2 illustrates the update flows. To receive signals from all edge devices, we utilize element-wise mean pooling for the RIS and the edge server. It should be noted that different functions can be chosen to ensure permutation equivariance and permutation invariance in aggregation and combining operations. In Sect. 5.6, we shall demonstrate the outstanding performance of our proposed design.

- *Parameter generation layer:* After going through D graphical mapping layers, the representation vectors are aggregated and combined to obtain comprehensive information. The final layer is the generation layer which maps these representations into the target transmit power, phase-shift vector, and denoising factor. For decoding the representation vectors of the RIS node, edge server node, and edge device nodes, we use three decoders, denoted by f_p, f_v, and f_η, respectively. Formally, we can express it as follows:

$$p_k = f_p(z_k^{(d)}), \eta = f_\eta(z_{K+1}^{(d)}), \boldsymbol{v} = f_v(z_{K+2}^{(d)}).$$

These decoders have the same design as the encoders in the initialization layer, and we utilize the Sigmoid activation layer to restrict their output range to $[-1, 1]$. By using an affine transformation, we can scale the corresponding design of each node to the necessary level.

The input and output dimension of all the learnable functions in the proposed GNN-based learning algorithms, i.e., $f_{\text{Comb}}^{d}(\cdot)$, $f_{\text{Comb},K+1}^{d}(\cdot)$, $f_{\text{Comb},K+2}^{d}(\cdot)$, $f^{d}(\cdot)$, $f_{K+1}^{d}(\cdot)$, $f_{K+2}^{d}(\cdot)$, $f_{\text{EC}}^{0}(\cdot)$, $f_p(\cdot)$, $f_v(\cdot)$, $f_\eta(\cdot)$, $\forall\, d = 0, \ldots, D$, are independent of the number of edge devices. Therefore, the proposed GNN-based learning algorithm is scalable and does not require to be re-trained when the number of edge devices varies, which is a key advantage of GNN over the conventional DNN methods.

5.5.3 *Loss Function Design and Training*

The proposed GNN-based algorithm is trained offline via mini-batch stochastic gradient descent methods in an unsupervised manner, and the training data in each mini-batch are uniformly sampled from the training set. To minimize the time-average error under the average transmit power constraints, the loss function is set as

$$\text{loss} = \frac{1}{B}\sum_{m=1}^{B}\left[\text{Er}(t) + \sum_{k\in\mathcal{K}}\text{Reg}(p_k(t), \bar{p}_k)\right], \tag{5.25}$$

where B denotes the size of mini-batch, and $\text{Reg}(p_k(t), \bar{p}_k)$ denotes the regularizer for the average power constraints. To ensure that the average power constraint is satisfied, we need an appropriate regularizer function $\psi(x, \bar{x})$, which should be 0 if x is less than or equal to $\bar{x}$ and a monotonic increasing function $h(x)$ otherwise. One way to simplify the computation during back-propagation is to set $h(x)$ as the identity function. In such cases, the ideal regularizer would be ReLU with the expression $\text{ReLU}(x - \bar{x}) = \max(0, x - \bar{x})$. By optimizing the variables that are coupled in the loss function, the average power constraints are fulfilled. To prevent overfitting, we halt the training process when the variance of the last 100 logarithmic training losses is less than a pre-defined threshold ϵ_{stop}.

5.6 Simulation Results

5.6.1 *Simulation Setup*

We consider an RIS-assisted over-the-air FEEL in a three-dimensional coordinate system, where an edge server and RIS are respectively deployed at $(-200, 10, 30)$ and $(0, 0, 10)$ meters, and the edge devices are uniformly distributed in the circular area centered at $(50, 0, 0)$ with a radius of 25 meters. The RIS reflection elements are arranged in a 10×10 pattern on the (y, z)-plane.

The channel response of the edge server-device k link is $h_k^{\text{d}}(t) = \gamma_k^{\text{d}}\tilde{h}_k^{\text{d}}(t)$, where $\gamma_k^{\text{d}} = \sqrt{D_0 d_{\text{DF},k}^{-\alpha_1}}$ denotes the large-scale fading coefficient, $\tilde{h}_k^{\text{d}}(t) \sim \mathcal{CN}(0, 1)$, $d_{\text{DF},k}$ is the distance between device k and the edge server, α_1 is the direct link's path loss exponent, and D_0 denotes the path loss at the reference distance. The channel coefficients $\boldsymbol{g}(t)$ and $\boldsymbol{h}_i^{\text{r}}(t)$ follow the Rician distribution, i.e.,

$$\boldsymbol{g}(t) = \sqrt{D_0 d_{\text{IF}}^{-\alpha_2}}\left(\sqrt{\frac{\delta}{\delta+1}}\tilde{\boldsymbol{g}}^{\text{LOS}}(t) + \sqrt{\frac{1}{1+\delta}}\tilde{\boldsymbol{g}}^{\text{NLOS}}(t)\right)$$

and

$$\boldsymbol{h}_k^{\mathrm{r}}(t) = \sqrt{D_0 d_{\mathrm{DR},k}^{-\alpha_3}} \left(\sqrt{\frac{\delta}{\delta+1}} \tilde{\boldsymbol{h}}_k^{\mathrm{r,LOS}}(t) + \sqrt{\frac{1}{1+\delta}} \tilde{\boldsymbol{h}}_k^{\mathrm{r,NLOS}}(t) \right),$$

where LOS and NLOS represent the line-of-sight and non-line-of-sight paths, respectively, d_{IF} and $d_{\mathrm{DR},k}$ are the distances between RIS and edge server, and between RIS and device k, respectively, and δ denotes the Rician factor. We set $\delta = 10$, $D_0 = -25$ dB, $\alpha_1 = 4$, $\alpha_2 = 1.8$ and $\alpha_3 = 2.1$. The non-line-of-sight paths, i.e., $\tilde{\boldsymbol{g}}^{\mathrm{NLOS}}(t)$ and $\tilde{\boldsymbol{h}}_k^{\mathrm{r,NLOS}}(t)$, obey the complex Gaussian distribution. For line-of-sight paths, by denoting the azimuth and elevation angles of departure (AOD) from the RIS to the edge server as θ_1^{A} and θ_1^{E}, respectively, we have $\tilde{\boldsymbol{g}}^{\mathrm{LOS}}(t) = \boldsymbol{a}_{\mathrm{RIS}}(\theta_1^{\mathrm{A}}, \theta_1^{\mathrm{E}}) a_{\mathrm{ES}}$, where $\boldsymbol{a}_{\mathrm{RIS}}(\theta_1^{\mathrm{A}}, \theta_1^{\mathrm{E}})$ denotes the steering vector of RIS, and a_{ES} denotes the steering scalar of edge server. In particular, $a_{\mathrm{ES}} = 1$ as one antenna is considered at the edge server. The n-th element of $\boldsymbol{a}_{\mathrm{RIS}}(\theta_1^{\mathrm{A}}, \theta_1^{\mathrm{E}})$ is $\left[\boldsymbol{a}_{\mathrm{RIS}}(\theta_1^{\mathrm{A}}, \theta_1^{\mathrm{E}})\right]_n = e^{j\beta \frac{2\pi d_{\mathrm{e}}}{\xi} \omega(\theta_1^{\mathrm{A}}, \theta_1^{\mathrm{E}}, n)}$, where $\omega(\theta_1^{\mathrm{A}}, \theta_1^{\mathrm{E}}, n) = \left\lfloor \frac{n-1}{10} \right\rfloor \sin(\theta_1^{\mathrm{E}}) + \mathrm{mod}(n-1, 10) \sin(\theta_1^{\mathrm{A}}) \cos(\theta_1^{\mathrm{E}})$, d_{e} denotes the interval between two adjacent passive elements of the RIS, and ξ denotes the carrier wavelength. Without loss of generality, we set $\frac{2\pi d_{\mathrm{e}}}{\xi} = 1$. With given locations of edge server $(x_{\mathrm{ES}}, y_{\mathrm{ES}}, z_{\mathrm{ES}})$ and RIS $(x_{\mathrm{RIS}}, y_{\mathrm{RIS}}, z_{\mathrm{RIS}})$, we have $\sin(\theta_1^{\mathrm{A}}) \cos(\theta_1^{\mathrm{E}}) = \frac{y_{\mathrm{ES}} - y_{\mathrm{RIS}}}{d_{\mathrm{IF}}}$ and $\sin(\theta_1^{\mathrm{E}}) = \frac{z_{\mathrm{ES}} - z_{\mathrm{RIS}}}{d_{\mathrm{IF}}}$. By denoting the angles of arrival (AOA) in the azimuth and elevation directions from device k to the RIS by $\theta_{2,k}^{\mathrm{A}}$ and $\theta_{2,k}^{\mathrm{E}}$, we have $\tilde{\boldsymbol{h}}_k^{\mathrm{r,LOS}}(t) = \boldsymbol{a}_{\mathrm{RIS}}(\theta_{2,k}^{\mathrm{A}}, \theta_{2,k}^{\mathrm{E}})$. With given location of device k (i.e., $(x_{\mathrm{d},k}, y_{\mathrm{d},k}, z_{\mathrm{d},k})$), we have $\sin(\theta_{2,k}^{\mathrm{A}}) \cos(\theta_{2,k}^{\mathrm{E}}) = \frac{y_{\mathrm{d},k} - y_{\mathrm{RIS}}}{d_{\mathrm{DR}}}$ and $\sin(\theta_{2,k}^{\mathrm{E}}) = \frac{z_{\mathrm{d},k} - z_{\mathrm{RIS}}}{d_{\mathrm{DR}}}$. The noise power is -75 dBm.

Datasets Our assessment involves testing the aforementioned algorithms on a handwritten digit recognition exercise utilizing the MNIST dataset. Equally divided data labeled with digits ranging from 0 to 9 is assigned to all edge devices. In this scenario, an fully-connected neural network is utilized for classification purposes, featuring three linear layers and employing the Sigmoid function as the activation function between adjacent linear layers. The loss function adopted for this experiment is cross-entropy.

5.6.2 Performance Comparison

The comparison of the training loss and test accuracy for various communication rounds is shown in Fig. 5.3. It can be observed from Fig. 5.3a that the proposed GNN-based learning algorithm achieves a much lower training loss than the optimization-based algorithm, especially after 40 rounds. It is because that the

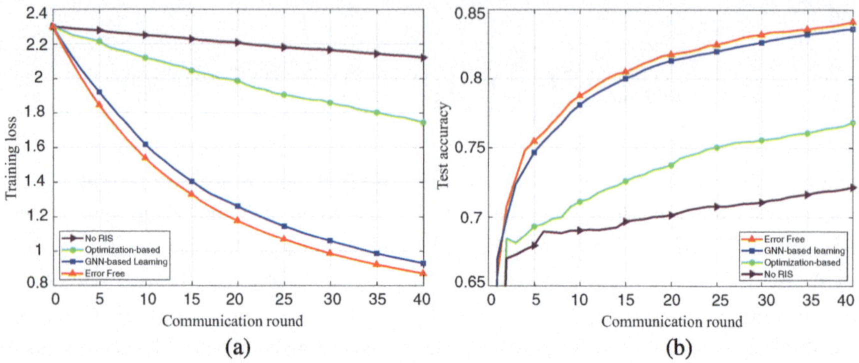

Fig. 5.3 Training loss and test accuracy versus number of communication rounds. (**a**) Training loss. (**b**) Test accuracy

proposed GNN-based algorithm can jointly optimize the AirComp transceiver and RIS phase-shifts by using unsupervised learning method with abundant samples. On the other hand, the optimization-based algorithm's alternating operation and convex relaxation result in a sub-optimal solution. Moreover, even the optimization-based method achieves a lower training loss than the scheme without RIS, which demonstrates the effectiveness of RIS in enhancing learning performance.

Figure 5.3b presents the test accuracy of all considered algorithms. It can be observed that the performance of all schemes increases with the increase of rounds. The proposed GNN-based learning algorithm achieves comparable accuracy to the Error Free scheme and outperforms the optimization-based and No RIS schemes. Moreover, as the same edge server and RIS locations are used for different sets of device locations, the permutation equivariance for both the edge server node and RIS node is ensured.

Figure 5.4 illustrates the impact of non-i.i.d. data on the test accuracy of the proposed GNN-based learning algorithm. Each edge device is assigned with the same amount of feature-label pairs that include two random categories. We can observe that the performance gap between the i.i.d. case and the non-i.i.d. case is slight. It is because that the deviation of the direction of the global gradient caused by the non-i.i.d. data is not significant when we consider only one local update. By balancing the tradeoff between signal-misalignment error and noise-induced error, our proposed GNN-based learning algorithm achieves a similar aggregation accuracy as the Error Free scheme. Due to the insufficient exploitation of RIS, the optimization-based algorithm and the No RIS scheme performs unsatisfactorily in terms of the test accuracy.

Figure 5.5 depicts how the number of RIS reflection elements influences the test accuracy. More reflection elements result in improved channel qualities, leading to better learning performance and more accurate model aggregation. When the number of reflection elements is increased from 5 to 120, the performance gap between the Error Free scheme and the proposed GNN-based algorithm gradually

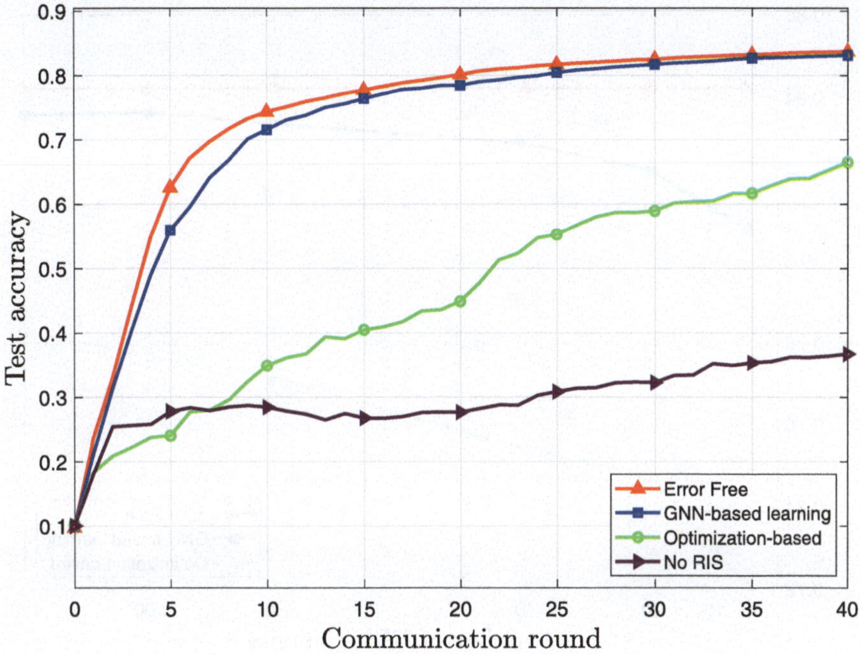

Fig. 5.4 Test accuracy versus number of communication rounds with non-i.i.d. data

get smaller. Although there is still a residual gap due to the initial optimality gap and gradient variance-induced gap defined in Theorem 5.1, the gradual performance improvement of proposed GNN-based algorithm confirms the effectiveness of RIS. Furthermore, the superiority of the proposed GNN-based learning algorithm over the optimization-based algorithm proves its ability to utilize RIS more efficiently for model aggregation.

5.7 Conclusions

In this chapter, we investigated an RIS-assisted over-the-air FEEL system. We theoretically derived the convergence upper bound of the proposed RIS-assisted over-the-air FEEL, and formulated a joint optimization problem with respect to the transmit power, denoising factor, and RIS phase-shifts. To reduce the computation complexity and enhance the learning performance, a GNN-based learning algorithm is developed to minimize the time-average error during the model aggregation. Finally, the superiority of the proposed GNN-based learning algorithm in terms of low-complexity, high training efficiency, and scalability are demonstrated by the extensive simulations.

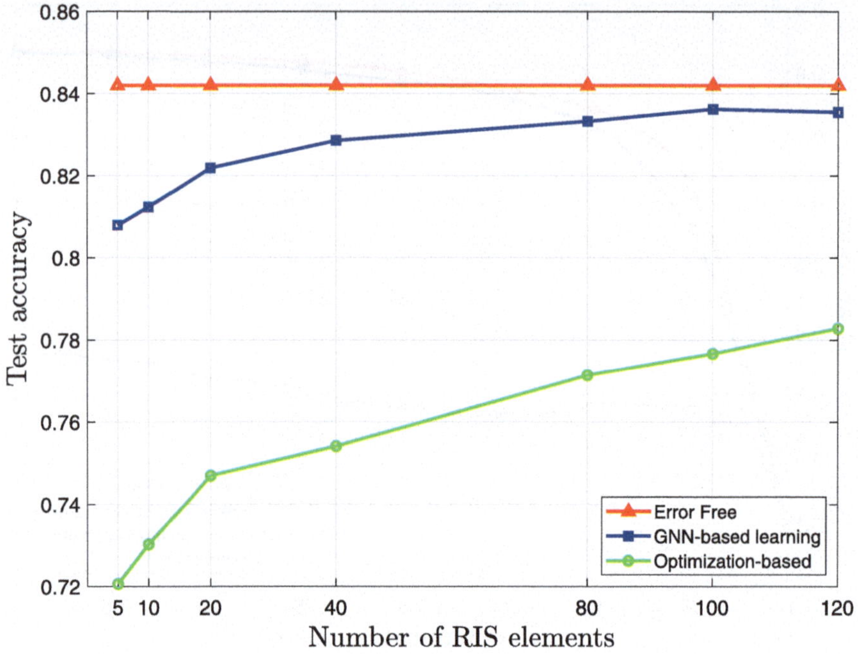

Fig. 5.5 Test accuracy versus number of reflection elements

References

1. X. Cao, G. Zhu, J. Xu, and K. Huang. Optimized power control for over-the-air computation in fading channels. *IEEE Trans. Wireless Commun.*, 19(11):7498–7513, Aug. 2020.
2. T. Chen, X. Zhang, M. You, G. Zheng, and S. Lambotharan. A GNN-based supervised learning framework for resource allocation in wireless IoT networks. *IEEE Internet Things J.*, 9(3):1712–1724, 2022.
3. M. Eisen and A. Ribeiro. Optimal wireless resource allocation with random edge graph neural networks. *IEEE Trans. Signal Process.*, 68:2977–2991, 2020.
4. W. Fang, M. Fu, Y. Shi, and Y. Zhou. Outage minimization for intelligent reflecting surface aided miso communication systems via stochastic beamforming. In *2020 IEEE 11th Sensor Array and Multichannel Signal Processing Workshop (SAM)*, pages 1–5. IEEE, 2020.
5. W. Fang, M. Fu, K. Wang, Y. Shi, and Y. Zhou. Stochastic beamforming for reconfigurable intelligent surface aided over-the-air computation. In *GLOBECOM 2020–2020 IEEE Global Communications Conference*, pages 1–6, 2020.
6. W. Fang, Y. Jiang, Y. Shi, Y. Zhou, W. Chen, and K. B. Letaief. Over-the-air computation via reconfigurable intelligent surface. *IEEE Transactions on Communications*, 69(12):8612–8626, 2021.
7. C. Huang, S. Hu, G. C. Alexandropoulos, A. Zappone, C. Yuen, R. Zhang, M. D. Renzo, and M. Debbah. Holographic MIMO surfaces for 6G wireless networks: Opportunities, challenges, and trends. *IEEE Wireless Commun.*, 27(5):118–125, May 2020.
8. C. Huang, R. Mo, and C. Yuen. Reconfigurable intelligent surface assisted multiuser MISO systems exploiting deep reinforcement learning. *IEEE J. Sel. Area. Commun.*, 38(8):1839–1850, 2020.

9. C. Huang, A. Zappone, G. C. Alexandropoulos, M. Debbah, and C. Yuen. Reconfigurable intelligent surfaces for energy efficiency in wireless communication. *IEEE Trans. Wireless Commun.*, 18(8):4157–4170, 2019.
10. C. B. Issaid, S. Samarakoon, M. Bennis, and H. V. Poor. Federated distributionally robust optimization for phase configuration of RISs. In *Proc. IEEE Global Commun. Conf. (GLOBECOM)*, pages 1–6, Madrid, Spain, Dec. 2021.
11. T. Jiang, H. V. Cheng, and W. Yu. Learning to reflect and to beamforming for intelligent reflecting surface with implicit channel estimation. *IEEE J. Sel. Areas Commun.*, 39(7):1931–1945, Jul. 2021.
12. H. Liu, X. Yuan, and Y. Zhang. Reconfigurable intelligent surface enabled federated learning: A unified communication-learning design approach. *IEEE Trans. Wireless Commun.*, 20(11):7595–7609, 2021.
13. S. Ma, W. Shen, X. Gao, and J. An. Robust channel estimation for RIS-aided millimeter-wave system with RIS blockage. *IEEE Trans. Veh. Technol.*, 71(5):5621–5626, 2022.
14. Z. Mao, M. Peng, and X. Liu. Channel estimation for reconfigurable intelligent surface assisted wireless communication systems in mobility scenarios. *China Commun.*, 18(3):29–38, 2021.
15. Y. Shen, Y. Shi, J. Zhang, and K. B. Letaief. Graph neural networks for scalable radio resource management: Architecture design and theoretical analysis. *IEEE J. Sel. Areas Commun.*, 39(1):101–115, Jan. 2021.
16. S. Wan, Z. Wang, and Y. Zhou. Scalable hybrid beamforming for multi-user miso systems: A graph neural network approach. *IEEE Trans. Wirel. Commun.*, pages 1–1, 2024.
17. Z. Wang, L. Liu, and S. Cui. Channel estimation for intelligent reflecting surface assisted multiuser communications: Framework, algorithms, and analysis. *IEEE Trans. Wireless Commun.*, 19(10):6607–6620, 2020.
18. Z. Wang, J. Qiu, Y. Zhou, Y. Shi, L. Fu, W. Chen, and K. Letaief. Federated learning via intelligent reflecting surface. *IEEE Trans. Wireless Commun*, 2021.
19. Z. Wang, Y. Zhou, Y. Zou, Q. An, Y. Shi, and M. Bennis. A graph neural network learning approach to optimize ris-assisted federated learning. *IEEE Trans. Wirel. Commun.*, 22(9):6092–6106, Sep. 2023.
20. K. Xu, W. Hu, J. Leskovec, and S. Jegelka. How powerful are graph neural networks? In *Proc. Int.l Conf. Learn. Represent. (ICLR)*, New Orleans, LA, May 2019.
21. K. Yang, Y. Shi, Y. Zhou, Z. Yang, L. Fu, and W. Chen. Federated machine learning for intelligent IoT via reconfigurable intelligent surface. *IEEE Network*, 34(5):16–22, Sept. 2020.
22. Y. Yang, Y. Zhou, Y. Wu, and Y. Shi. Differentially private federated learning via reconfigurable intelligent surface. *arXiv:2203.17028*, 2022.
23. W. Yu and R. Lui. Dual methods for nonconvex spectrum optimization of multicarrier systems. *IEEE Trans. Communi.*, 54(7):1310–1322, 2006.
24. G. Zhou, C. Pan, H. Ren, P. Popovski, and A. L. Swindlehurst. Channel estimation for RIS-aided multiuser millimeter-wave systems. *IEEE Trans. Signal Process.*, 70:1478–1492, 2022.

Chapter 6
Federated Edge Learning via Unmanned Aerial Vehicle

6.1 Overview

The primary hurdle to implementing FEEL on wireless networks stems from communication obstacles in the form of stragglers who have weak links [8, 9]. A case in point is digital synchronous FEEL [2], where stragglers have a notable impact on slowing down the model aggregation process and leading to extended cumulative communication delay, since the parameter server (PS) has to wait for updates from all the participants. With a rise in the number of stragglers, the communication delay becomes intolerable. It follows that addressing the straggler issue holds the key to developing communication-efficient FEEL systems.

Numerous attempts have been made to mitigate the straggler effect in FEEL, including device scheduling techniques [2, 16, 20]. For example, to minimize model misalignment error caused by stragglers in AirComp-based FEEL, researchers in [16, 20] scheduled devices with reliable communication channels for simultaneous model uploading. Similarly, in digital FEEL, devices with large contributions to the global model [2] or favorable channel conditions [1, 12] are usually chosen to decrease the communication delay incurred by stragglers. However, these approaches necessitate a terrestrial BS for network coverage during model aggregation, which may not always be feasible in remote areas where terrestrial networks are unavailable. Consequently, it is imperative to deploy a more adaptable PS that can proactively establish optimal communication links among devices.

Unmanned aerial vehicles (UAVs) present a promising and supplementary alternative to terrestrial networks [10, 13, 14, 17], enabling seamless connectivity and coverage extension to support diverse FEEL tasks. In [14], a multi-UAV cooperative communication framework is proposed to reduce access delay. In [17], the UAVs are deployed in a wireless network to collect data. In [10], a UAV is utilized as an edge server to provide computing service for mobile users. Thanks to the flexibility of UAVs, the straggler effect of FEEL can be alleviated by establishing

Y. Zhou et al., *Federated Edge Learning*, Wireless Networks,
https://doi.org/10.1007/978-3-031-96649-1_6

short-distance line of sight links to scheduled devices, thereby reducing cumulative model aggregation loss and accelerating FEEL convergence.

In this chapter, we examine a UAV-based FEEL network where a flying PS in the form of a UAV is deployed to aggregate and update digital FEEL model parameters in areas lacking terrestrial BS coverage [5]. However, as the flight time of the UAV is limited, it is a challenge to balance the tradeoff between completion time and convergence accuracy. To minimize the completion time while considering the convergence accuracy and address the communication straggler issues that may arise in UAV-enabled FEEL networks, we propose a joint design approach for device scheduling and UAV trajectory planning. We analyze the convergence of the FEEL algorithm, followed by developing a unified optimization framework based on the principle of block coordinate descent and Lagrange duality.

6.2 System Model

This chapter focuses on a UAV-assisted FEEL network, as depicted in Fig. 6.1. The network comprises a single-antenna UAV and a set $\mathcal{K}$ of K single-antenna devices that aim to collaboratively learn an ML model. To cater to the sparse deployment and unavailability of terrestrial BSs in underdeveloped areas, we employ a UAV with a potent processor to act as a PS in these harsh circumstances. Each device k possesses a local dataset $\mathcal{D}_k$ that contains D_k data samples. The local dataset $\mathcal{D}_k$ is made up of data samples $\{\boldsymbol{x}_{ki}, y_{ki}\}_{i=1}^{D_k}$ where $\boldsymbol{x}_{ki} \in \mathbb{R}^d$ is the input sample vector containing d features, and $y_{ki} \in \mathbb{R}$ is the label for sample $\boldsymbol{x}_{ki}$. The UAV-enabled FEEL system's primary objective is to cooperatively learn an ML model while keeping the local training data on mobile devices under the coordination of the UAV in an iterative manner.

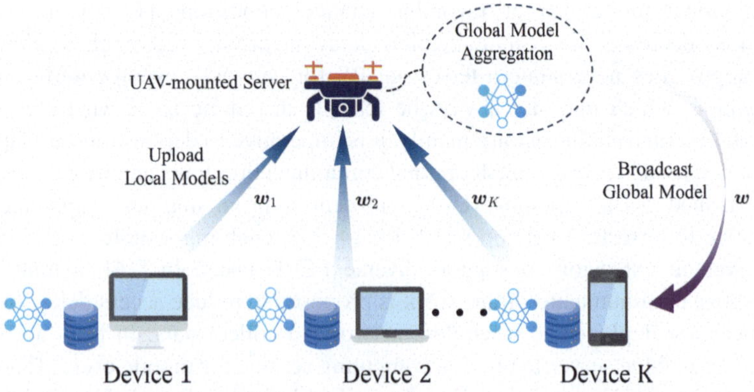

Fig. 6.1 FEEL over UAV-enabled wireless networks

6.2.1 FEEL Training Model

The local loss function of device k is defined as

$$\mathcal{F}_k(\boldsymbol{w}) = \frac{1}{D_k}\sum_{i=1}^{D_k} f(\boldsymbol{w}; \boldsymbol{x}_{ki}, y_{ki}), \forall k \in \mathcal{K}, \tag{6.1}$$

where $\boldsymbol{w} \in \mathbb{R}^d$ and $\boldsymbol{w}_k \in \mathbb{R}^d$ denote the global model at the UAV and the local model at the k-th device, respectively. Accordingly, the global loss function at the UAV is given by

$$\mathcal{F}(\boldsymbol{w}) := \frac{1}{D}\sum_{k=1}^{K} D_k\mathcal{F}_k(\boldsymbol{w}), \tag{6.2}$$

where D denotes the total size of data with $D = \sum_{k=1}^{K} D_k$. Mathematically, FEEL aims to solve the following problem:

$$\begin{aligned} &\underset{\boldsymbol{w}_1,\ldots,\boldsymbol{w}_K}{\text{minimize}} \quad \frac{1}{D}\sum_{k=1}^{K}\sum_{i=1}^{D_k} f(\boldsymbol{w}_k; \boldsymbol{x}_{ki}, y_{ki}) \\ &\text{subject to} \quad \boldsymbol{w}_1 = \cdots = \boldsymbol{w}_K = \boldsymbol{w}, \end{aligned} \tag{6.3}$$

where constraint (6.3) is used to ensure that all devices come to a consensus. To accomplish this goal, the server must frequently communicate with edge devices. During a communication round, the devices first update their local models based on their own data and a common model initialization from the parameter server, and then upload the model updates to the server via wireless channels, as outlined below.

To begin with, the UAV identifies which devices will participate in the ongoing round. Using the binary variable $a_k[n] \in \{0, 1\}$ to represent the device scheduling for the n-th round, where $a_k[n] = 1$ indicates that device k will transmit its updated local model to the UAV during communication round n, and $a_k[n] = 0$ implies that it will not. We define $\mathcal{A}_n$ as the set of devices scheduled to participate in round n. NSubsequently, the UAV broadcasts the current global model, denoted by $\boldsymbol{w}_{n-1}$ to all devices within the scheduled set.

Each scheduled device $k \in \mathcal{A}_n$ receives the global model and updates its local model by gradient descent based on its local dataset:

$$\boldsymbol{w}_{k,n} = \boldsymbol{w}_{n-1} - \frac{\eta}{D_k}\sum_{(x_{ki},y_{ki})\in\mathcal{D}_k} \nabla f(\boldsymbol{w}_{n-1}; \boldsymbol{x}_{ki}, y_{ki}), \tag{6.4}$$

where η is the learning rate and $\nabla f(\boldsymbol{w}_{n-1}; \boldsymbol{x}_{ki}, y_{ki})$ is the gradient of $f(\boldsymbol{w}; \boldsymbol{x}_{ki}, y_{ki})$ with respect to $\boldsymbol{w}$ at $\boldsymbol{w}_{n-1}$.

Once all devices scheduled for a round have uploaded their local models, the UAV combines these updates to generate an updated global model using the following formula:

$$\boldsymbol{w}_n = \frac{\sum_{k=1}^{K} a_k[n]\boldsymbol{w}_{k,n}}{\sum_{k=1}^{K} a_k[n]}, \tag{6.5}$$

where $\boldsymbol{w}_n$ denotes the new global model generated at the UAV in the n-th communication round. By using Eqs. (6.4) and (6.5), both the UAV and devices can continuously update their models until they reach the maximum number of communication rounds allowed. Since the devices update local FEEL models and transmit them via wireless cellular links, device scheduling significantly impacts both learning performance and communication delays. Additionally, due to the UAV's mobility, the channel conditions between the UAV and devices may vary across different time slots. Consequently, optimizing device scheduling, time allocation, and UAV trajectory is essential to accommodate time-varying channel conditions and achieve the desired FEEL convergence accuracy. In the next subsection, we will provide further details about the communication and computation models used in FEEL over UAV-enabled networks.

6.2.2 UAV-Enabled Signal Transmission

We assume that all ground devices are stationary, and their locations are known to the UAV. The horizontal position of device k is represented as $\mathbf{u}_k = [\tilde{x}_k, \tilde{y}_k] \in \mathbb{R}^{1\times 2}$, where $\tilde{x}_k$ and $\tilde{y}_k$ denote its x- and y-coordinates, respectively. We consider a fixed number of communication rounds N and focus on the time required for the UAV to complete the FEEL process, denoted as T. While energy efficiency is an important design consideration in UAV-enabled systems [11], we do not address this issue in our current study, leaving it for future work. However, we note that optimizing completion time may improve the energy efficiency of rotary-wing UAVs, especially when hovering. The location of the UAV in a three-dimensional Cartesian coordinate system can be projected onto the horizontal plane as $\boldsymbol{q}(t) = [x(t), y(t)] \in \mathbb{R}^{1\times 2}$ for $0 \le t \le T$, where $x(t)$ and $y(t)$ represent its x- and y-coordinates at time instant t, respectively. We assume that the UAV maintains a constant altitude H above ground level throughout the FEEL process and begins its mission from an initial location marked as $\boldsymbol{q}_I = [x_I, y_I] \in \mathbb{R}^{1\times 2}$. In practice, the minimum altitude H must be sufficient for obstacle avoidance without the need for frequent ascent and descent, while the initial location is determined by various factors such as energy replenishment [18]. Therefore, we have $\boldsymbol{q}(0) = \boldsymbol{q}_I$. Additionally, the maximum speed of the UAV is denoted as $V_{\max}$ in meters per second (m/s), and its velocity can vary over time under the constraint that $\sqrt{\dot{x}^2(t) + \dot{y}^2(t)} \le V_{\max}$ for $0 < t < T$, where $\dot{x}(t)$ and $\dot{y}(t)$ represent the time derivatives of $x(t)$ and $y(t)$, respectively.

To enable UAV trajectory planning, we employ the time discretization method commonly utilized in previous studies [6, 7, 18] to handle the continuous UAV trajectory. In this approach, we divide the total mission duration T into N unequal time slots, with the n-th slot having a duration of $\delta[n]$ and $\sum_{n=1}^{N} \delta[n] = T$, where $\delta[n]$ denotes the sum of the computation time, model uploading time, and model downloading time for round n. Furthermore, to ensure effective time discretization, we assume that the distance covered by the UAV within each time slot does not exceed a maximum distance $\Delta q_{\max}$, which satisfies $\Delta q_{\max} \ll H$ to maintain a relatively constant distance between the UAV and devices in each slot. Based on this technique, the UAV's trajectory over time T can be approximated using a sequence of $(N+1)$ coordinates denoted as $\{\boldsymbol{q}[n]\}_{n=0}^{N}$, where $\boldsymbol{q}[n]$ represents the UAV's horizontal position at time slot n. The UAV's mobility constraints are given by

$$\|\boldsymbol{q}[n]-\boldsymbol{q}[n-1]\|_2 \leq \min\{V_{\max}\delta[n], \Delta q_{\max}\}, n = 1, \ldots, N, \tag{6.6}$$

$$\boldsymbol{q}[0] = \boldsymbol{q}_I, \tag{6.7}$$

where (6.6) and (6.7) correspond to the UAV's speed constraint and its initial location constraint, respectively. Additionally, we assume that the communication link between each device and the UAV is dominated by the line-of-sight (LoS) channel due to the high altitude at which UAVs typically operate, consistent with existing literature on UAVs [6, 7, 19]. Using the free-space path loss model [11], the channel gain between device k and the UAV at time slot n is given by

$$h_k[n] = \beta_0 d_k^{-2}[n] = \frac{\beta_0}{H^2 + \|\boldsymbol{q}[n] - \mathbf{u}_k\|_2^2}, \tag{6.8}$$

where β_0 represents the channel gain at the reference distance of 1 meter, and $d_k[n]$ denotes the distance between device k and the UAV at time slot n.

We describe the process of FEEL over UAV-enabled wireless transmission during each time slot $\delta[n]$, which comprises four stages: local computation, local model uploading, global computation, and global model downloading, as depicted in Fig. 6.2.

6.2.2.1 Local Computation

The number of required processing cycles for computing one data sample at device k is denoted by c_k, which can be determined offline and is known a priori. As all samples $\{\boldsymbol{x}_{ki}, y_{ki}\}_{i\in\mathcal{D}_k}$ have the same size (i.e., number of bits), the number of CPU cycles needed for device k to execute one local round is $c_k D_k$. By letting the CPU-cycle frequency of device k be denoted by f_k, the computation time per round for device k is given by

$$t_k^{\text{comp}} = \frac{c_k D_k}{f_k}. \tag{6.9}$$

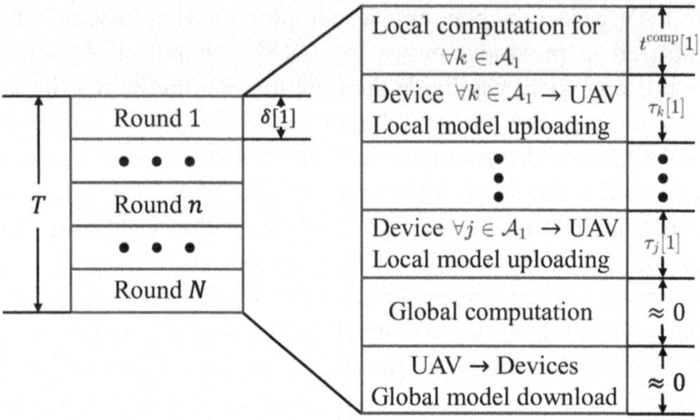

Fig. 6.2 The illustration of UAV-enabled signal transmission

The CPU energy consumption of device k for the n-th round of local computation is [4]

$$E_k^{\text{comp}}[n] = a_k[n] \sum_{i=1}^{c_k D_k} \frac{\alpha_k}{2} f_k^2 = a_k[n] \frac{\alpha_k}{2} c_k D_k f_k^2, \tag{6.10}$$

where $\alpha_k/2$ represents the effective capacitance coefficient of device k's computing chipset. Both synchronous and asynchronous FEEL schemes are utilized for updating the global parameters, although the former is more commonly used due to its simplicity and guaranteed FEEL convergence capability for resource-constrained devices. Therefore, we adopt the synchronous FEEL aggregation technique in this chapter as in most existing FEEL studies [2–4, 12]. Specifically, we consider synchronous operation that requires all scheduled devices to simultaneously train their local models and complete their training before entering the communication phase. Thus, the time required for local model training in the n-th round is given by [4]

$$t^{\text{comp}}[n] = \max_k \{a_k[n] t_k^{\text{comp}}\}. \tag{6.11}$$

Therefore, the computation time in one local round is determined by the scheduled devices with large data sizes and low CPU frequencies.

6.2.2.2 Local Model Uploading

The data rate (bit/s) that can be achieved by device k at the n-th round is determined by parameters such as the system bandwidth (B), the transmit power ($p_k[n]$) of the device, and the power of the additional complex Gaussian noise (σ^2), according to Eq. (6.12). Here, $\tau_k[n]$ denotes the time duration for device k to transmit its local model to the UAV at the n-th round. It is worth noting that the dimension of the model parameter vector (s in bits) remains constant throughout the FEEL process. The upload time for device k can be calculated using Eq. (6.13). In general, a higher channel gain leads to shorter uploading times for a given data size, and the UAV can optimize the uploading time of scheduled devices by establishing LoS connections and shortening communication distances with the devices. In terms of energy consumption, Eq. (6.14) indicates that device k consumes energy proportional to the duration for transmitting the local model ($\tau_k[n]$), the power of the additional complex Gaussian noise (σ^2), and the system bandwidth (B).

To ensure smooth uploading of local models without interference, a TDMA scheme is utilized during the local model uploading stage as illustrated in Fig. 6.2. During slot n, each of the $|\mathcal{A}_n|$ devices transmits its local model one by one. The achievable data rate (bit/s) of device k at the n-th round is defined as

$$r_k[n] = B \log_2 \left(1 + \frac{p_k[n]h_k[n]}{\sigma^2}\right), \tag{6.12}$$

where B is the system bandwidth in Hertz, $p_k[n]$ is the transmit power of device k at the n-th round, and σ^2 is the power of the additional complex Gaussian noise. Denote the time taken by device k to transmit its local model to the UAV during round n of federated learning as $\tau_k[n]$. It is important to note that the dimension of the model parameter vector, measured in bits and denoted by s, remains constant throughout the FEEL process. The uploading time of device k can be calculated as

$$\tau_k[n] = \frac{a_k[n]s}{r_k[n]}. \tag{6.13}$$

If we consider a fixed data size, it can be observed that an increase in channel gain results in a decrease in the duration required for uploading. The UAV possesses the ability to reduce the upload time for scheduled devices through various means, such as establishing line-of-sight connections and minimizing communication distances with the devices. By combining (6.12) and (6.13), to transmit s within a time duration $\tau_k[n]$, the device k's energy consumption is

$$E_k^{\text{comm}}[n] = \frac{\tau_k[n]\sigma^2}{h_k[n]} \left(2^{\frac{a_k[n]s}{B\tau_k[n]}} - 1\right). \tag{6.14}$$

Upon completion of the local training model uploads by all scheduled devices, denoted as $\{\boldsymbol{w}_{k,n}\}$ at slot n, the UAV proceeds to perform a global update of the model. The resultant updated model, which is of length s bits, is then broadcasted to all connected devices. We use the variables f_0, c_0, and P_0 to represent the fixed computing speed of the UAV's processor, the required number of processing cycles for one local model at the UAV, and the transmit power of the UAV, respectively. The time that spent on updating the global model and feeding the result back to devices at the UAV is expressed as

$$t^{\mathrm{UAV}}[n] = \frac{K c_0}{f_0} + \frac{s}{\min_{k \in \mathcal{A}_n} B \log_2 \left(1 + \frac{P_0 h_k[n]}{\sigma^2}\right)}. \tag{6.15}$$

Compared to the time taken for local model computation at device k, which is represented by $\frac{D_k c_k}{f_k}$ in Eq. (6.15), the first term can be ignored due to the superior computational capabilities of the UAV (f_0) over devices (f_k) [19]. Additionally, since the number of devices (K) is significantly lower than the size of the local dataset (D_k), the first term can be safely neglected. On another note, the UAV only needs to broadcast the global model to the scheduled devices while each device uploads their local models using TDMA. As a result, we can conclude that $t^{\mathrm{UAV}}[n] \ll \sum_{k=1}^{K} \tau_k[n] + t^{\mathrm{comp}}[n]$ as stated in [4, 19]. Consequently, the total uploading time for the devices will take up the remaining time after local model training at the n-th round, i.e., $\sum_{k=1}^{K} \tau_k[n] \leq \delta[n] - t^{\mathrm{comp}}[n]$, $\forall n$.

6.3 Convergence Analysis and Problem Formulation

Next, we present the convergence analysis of UAV-enabled FEEL. To be specific, we characterize the impact of system parameters on the convergence performance. Based on the analysis, we formulate a training time minimization problem to improve the training efficiency.

6.3.1 Convergence Analysis of FEEL

Our analysis relies on the following assumptions, which are standard in non-convex optimization.

Assumption 12 (Smoothness) *The function $\mathcal{F}$: $\mathbb{R}^d \to \mathbb{R}$ is L-smooth. That is, $\forall \boldsymbol{w}, \boldsymbol{w}' \in \mathbb{R}^d$,*

$$\mathcal{F}(\boldsymbol{w}) \leq \mathcal{F}(\boldsymbol{w}') + \langle \nabla \mathcal{F}(\boldsymbol{w}'), \boldsymbol{w} - \boldsymbol{w}' \rangle + \frac{L}{2} \|\boldsymbol{w}' - \boldsymbol{w}\|_2^2. \tag{6.16}$$

Assumption 13 (Bounded Loss Function) *For any parameter vector $\boldsymbol{w}$, the loss function $\mathcal{F}(\boldsymbol{w})$ is lower-bounded by $F^\star$.*

Assumption 14 (Bounded Sample-Wise Gradient Norm) *For a constant $\kappa > 0$, the sample-wise gradients at local devices are bounded*

$$\|\nabla f(\boldsymbol{w}_n; \boldsymbol{x}, y)\|_2^2 \leq \kappa. \tag{6.17}$$

Based on the above assumptions, the convergence behavior of our considered algorithm is satisfied.

Theorem 6.1 *Suppose that the loss functions satisfy assumptions 1–3, given the learning rate* $0 \leq \eta \leq \frac{1}{L}$*, after N rounds, the average norm of the global gradients is upper bounded by*

$$\frac{1}{N}\sum_{n=0}^{N-1} \|\nabla \mathcal{F}(\boldsymbol{w}_n)\|^2 \leq \frac{2}{N\eta}(\mathcal{F}(\boldsymbol{w}_0) - F^{\star})$$
$$+\frac{4K\kappa}{ND^2}\sum_{n=0}^{N-1}\sum_{k=1}^{K}(1 - a_k[n+1])D_k^2. \tag{6.18}$$

The second term in (6.18) represents the time-average aggregation error caused by device scheduling, which decreases as the number of scheduled devices increases. Although scheduling a large number of devices for simultaneous model uploading might seem beneficial, it can significantly increase the communication burden and lead to substantial delays during the model aggregation process. These delays can be exacerbated if some devices encounter communication stragglers. Therefore, it is crucial to carefully select an appropriate subset of devices to balance training performance and time consumption, ensuring optimal results.

6.3.2 Problem Formulation

Let $\boldsymbol{A} = \{a_k[n], \forall n \in \mathcal{N}, \forall k \in \mathcal{K}\}$, $\boldsymbol{\Gamma} = \{\tau_k[n], \forall n \in \mathcal{N}, \forall k \in \mathcal{K}\}$, $\boldsymbol{Q} = \{\boldsymbol{q}[n], \forall n = 0, \ldots, N\}$, and $\boldsymbol{\delta} = \{\delta[n], \forall n = 1, \ldots, N\}$. In this chapter, given the number of communication rounds N, we aim to minimize the completion time of the FEEL training process under the target convergence accuracy requirements by jointly optimizing the devices' scheduling variables $\boldsymbol{A}$, time allocation $\boldsymbol{\Gamma}$, UAV trajectory $\boldsymbol{Q}$, and slot intervals $\boldsymbol{\delta}$. The corresponding optimization problem is formulated as

$$\underset{\boldsymbol{A},\boldsymbol{\Gamma},\boldsymbol{Q},\boldsymbol{\delta}}{\text{minimize}} \sum_{n=1}^{N}\delta[n]$$

$$\text{subject to } a_k[n] \in \{0, 1\}, \forall k \in \mathcal{K}, \forall n \in \mathcal{N}, \tag{6.19a}$$

$$\sum_{n=1}^{N}\left(E_k^{\text{comm}}[n] + E_k^{\text{comp}}[n]\right) \leq E_k, \forall k \in \mathcal{K}, \tag{6.19b}$$

$$\sum_{k=1}^{K}\tau_k[n] + \max_k a_k[n]t_k^{\text{comp}} \leq \delta[n], \forall n \in \mathcal{N}, \tag{6.19c}$$

$$\frac{2}{N\eta}(\mathcal{F}(\boldsymbol{w}_0) - F^{\star}) + \frac{4K\kappa}{ND^2}\sum_{n=0}^{N-1}\sum_{k=1}^{K}(1 - a_k[n+1])D_k^2 \leq \epsilon, \quad (6.19\text{d})$$

$$\text{Constraints (6.6), (6.7).} \quad (6.19\text{e})$$

The constraints in (6.19a) pertain to the integer nature of the devices' scheduling variables. The energy budgets of the devices are represented by the constraints in (6.19b). Constraints (6.19c) ensure that communication and computation time per round do not exceed the duration of each time slot. The target convergence accuracy requirement of FEEL is denoted by constraint (6.19d). Problem (6.19) presents challenges due to several reasons. Firstly, the optimization variables $\boldsymbol{A}$ for device scheduling are binary, leading to integer constraints in (6.19a)–(6.19d). Secondly, $E_k^{\text{comm}}[n]$ in constraints (6.19b) are not jointly convex with respect to the optimization variables $\boldsymbol{A}$, $\boldsymbol{\Gamma}$, $\boldsymbol{Q}$, and $\boldsymbol{\delta}$. As a result, problem (6.19) is a mixed-integer non-convex problem that is generally difficult to solve optimally.

6.3.3 Problem Analysis

Before addressing problem (6.19), it is important to verify its feasibility. As stated in (6.18), achieving the minimum convergence bound requires scheduling all devices in every communication round. Thus, if the minimum value of the convergence bound exceeds the desired target accuracy, then problem (6.19) cannot be solved. Additionally, if E_k values for all k are insufficient, then the constraints outlined in (6.19b) may not be met, even with the lowest energy consumption from scheduled devices. In summary, we propose the following approach to assess the feasibility of problem (6.19).

Proposition 6.1 *Problem* (6.19) *is feasible if and only if*

$$N \geq \lceil \frac{2(\mathcal{F}(\boldsymbol{w}_0) - F^{\star})}{\epsilon\eta} \rceil \text{ and } \sum_{k=1}^{K} E_k \geq |S|\frac{\sigma^2 s H^2 \ln 2}{\beta_0 B}, \quad (6.20)$$

where $|S|$ *denotes the required amount of scheduled devices during training process, given by* $|S| = \lceil KN - (\epsilon - \frac{2}{N\eta}(\mathcal{F}(\boldsymbol{w}_0) - F^{\star}))\frac{ND^2}{4K\kappa \max_k D_k^2} \rceil$.

By using Proposition 6.1, we can check whether Problem (6.19) is feasible.

6.4 Device Scheduling and Time Allocation Along with Trajectory Design

The duration for updating the local models is determined by the device scheduling matrix $\boldsymbol{A}$ and UAV trajectory $\boldsymbol{Q}$, as seen in (6.19). With the aid of the UAV's mobility, we can establish preferable connections with devices that are scheduled in each round. This approach helps to address the communication straggler problem, which ultimately leads to a reduction in the FEEL completion time. In this section, we propose the BCD-LD method, which involves designing the device scheduling, time allocation, and UAV trajectory. Our methodology begins by optimizing the device scheduling and time allocation for a given UAV trajectory. Subsequently, we optimize the UAV trajectory $\boldsymbol{Q}$ after deriving the optimal and closed-form solutions for each optimization subproblem using LD methods. By adopting these measures, we can mitigate the high computational complexity associated with the optimization process.

6.4.1 *Joint Device Scheduling and Time Allocation*

To eliminate the max function in constraints (6.19c), it is equivalently converted into

$$\sum_{k=1}^{K} \tau_k[n] + a_k[n]t_k^{\text{comp}} \leq \delta[n], \forall k, \forall n \in \mathcal{N}. \tag{6.21}$$

Given trajectory $\boldsymbol{Q}$, for constraints (6.19d) and (6.6), they can be respectively rewritten as

$$\sum_{n=0}^{N-1}\sum_{k=1}^{K} a_k[n+1]D_k^2 \geq C, \tag{6.22}$$

$$\delta[n] \geq \|\boldsymbol{q}[n] - \boldsymbol{q}[n-1]\|_2/V_{\max}, \forall n. \tag{6.23}$$

where $C = N\sum_{k=1}^{K} D_k^2 - (\epsilon - \frac{2}{N\eta}(\mathcal{F}(\boldsymbol{w}_0) - F^{\star}))\frac{ND^2}{4K\kappa}$ is constant. By further relaxing the integer constraints, i.e., (6.19a), Problem (6.19) under fixed UAV trajectory $\boldsymbol{Q}$ reduces to the following problem,

$$\underset{\boldsymbol{A},\boldsymbol{\delta},\boldsymbol{\tau}}{\text{minimize}} \sum_{n=1}^{N} \delta[n]$$

$$\text{subject to} \sum_{n=1}^{N}\left(\frac{\tau_k[n]\sigma^2}{h_k[n]}\left(2^{\frac{a_k[n]s}{B\tau_k[n]}} - 1\right) + a_k[n]E_k^{\text{comp}}[n]\right) \leq E_k, \forall k \in \mathcal{K},$$

$$0 \le a_k[n] \le 1, \forall k \in \mathcal{K}, \forall n \in \mathcal{N}, \quad (6.24)$$

Constraints (6.21), (6.22), and (6.23).

Define $\frac{\tau_k[n]\sigma^2}{h_k[n]}\left(2^{\frac{a_k[n]s}{B\tau_k[n]}} - 1\right) = 0$ when $\tau_k[n] = 0, \forall k, \forall n$, such that the left-hand-sides of constraints (6.24) are continuous with respect to $\tau_k[n]$ with $\tau_k[n] \ge 0$. Accordingly, since $\frac{\sigma^2}{h_k[n]}(2^{\frac{a_k[n]s}{B}} - 1)$ is convex with respect to $a_k[n]$, its perspective function $\frac{\tau_k[n]\sigma^2}{h_k[n]}\left(2^{\frac{a_k[n]s}{B\tau_k[n]}} - 1\right)$ with $\tau_k[n] \ge 0$ in constraints (6.24) is jointly convex with respect to $a_k[n]$ and $\tau_k[n]$. Furthermore, all remaining constraints in problem (6.24) are affine constraints. As such, it is concluded that problem (6.24) is convex, which can be optimally solved by the Lagrange duality method. The partial Lagrange function of problem (6.24) is given by

$$\begin{aligned}
&\mathcal{L}(\boldsymbol{\delta}, \boldsymbol{A}, \boldsymbol{\tau}, \boldsymbol{\lambda}, \boldsymbol{\mu}, \xi) \\
&= \sum_{n=1}^{N} \delta[n] + \sum_{k=1}^{K} \lambda_k \Big(\sum_{n=1}^{N} \Big(\frac{\tau_k[n]\sigma^2}{h_k[n]} \Big(2^{\frac{a_k[n]s}{B\tau_k[n]}} - 1\Big) + a_k[n] E_k^{\text{comp}}[n] \Big) - E_k \Big) \\
&+ \sum_{n=1}^{N} \sum_{k=1}^{K} \mu_k[n] \Big(\sum_{k=1}^{K} \tau_k[n] + a_k[n] t_k^{\text{comp}} - \delta[n] \Big) \\
&+ \xi \Big(C - \sum_{n=0}^{N-1} \sum_{k=1}^{K} a_k[n+1] D_k^2\Big),
\end{aligned}$$

where $\boldsymbol{\lambda} = \{\lambda_k, \forall k\}$, $\boldsymbol{\mu} = \{\mu_k[n], \forall k, \forall n\}$, and ξ are the non-negative Lagrange multipliers associated with constraints (6.24), (6.21), and (6.22), respectively. The other boundary constraints in problem (6.24) will be absorbed into the optimal solution in the following. Accordingly, the dual function is given by

$$\begin{aligned}
g(\boldsymbol{\lambda}, \boldsymbol{\mu}, \xi) = \underset{\boldsymbol{\delta}, \boldsymbol{A}, \boldsymbol{\tau}, \xi}{\text{minimize}} \ & \mathcal{L}(\boldsymbol{\delta}, \boldsymbol{A}, \boldsymbol{\tau}, \boldsymbol{\lambda}, \boldsymbol{\mu}) \\
\text{subject to} \ & \text{Constraints (6.24) and (6.21).} \qquad (6.25)
\end{aligned}$$

To make $g(\boldsymbol{\lambda}, \boldsymbol{\mu}, \xi)$ lower-bounded (i.e., $g(\boldsymbol{\lambda}, \boldsymbol{\mu}, \xi) > -\infty$), the condition $\sum_{k=1}^{K} \mu_k[n] \le 1$ must hold. Therefore, the dual problem of problem (6.24) is given by

$$\begin{aligned}
\underset{\boldsymbol{\lambda}, \boldsymbol{\mu}, \xi}{\text{maximize}} \ & g(\boldsymbol{\lambda}, \boldsymbol{\mu}, \xi) \\
\text{subject to} \ & \sum_{k=1}^{K} \mu_k[n] > 1, \forall n, \qquad (6.26a) \\
& \boldsymbol{\lambda} \succeq 0, \boldsymbol{\mu} \succeq 0, \xi \ge 0. \qquad (6.26b)
\end{aligned}$$

Since the strong duality holds, we can solve problem (6.24) by equivalently solving its dual problem (6.26). In the following, we show how to obtain the dual function $g(\boldsymbol{\lambda}, \boldsymbol{\mu}, \xi)$ by solving problem (6.25) under any given $\boldsymbol{\lambda}, \boldsymbol{\mu}, \xi$ that satisfy (6.26a) and (6.26b). By ignoring the constant terms, problem (6.24) reduces to the following formulation,

$$\begin{aligned}
\underset{\boldsymbol{\delta},\boldsymbol{A},\boldsymbol{\tau}}{\text{minimize}} \quad & \sum_{n=1}^{N}\Big(1-\sum_{k=1}^{K}\mu_k[n]\Big)\delta[n]+\sum_{k=1}^{K}\lambda_k\sum_{n=1}^{N}\Big(\frac{\tau_k[n]\sigma^2}{h_k[n]}\big(2^{\frac{a_k[n]s}{B\tau_k[n]}}-1\big) \\
& +a_k[n]E_k^{\text{comp}}[n]\Big)+\sum_{n=1}^{N}\sum_{k=1}^{K}\mu_k[n]\Big(\sum_{k=1}^{K}\tau_k[n]+a_k[n]t_k^{\text{comp}}\Big) \\
& -\xi\sum_{n=1}^{N}\sum_{k=1}^{K}a_k[n]D_k^2 \\
\text{subject to} \quad & \text{Constraints (6.24) and (6.21)}, \qquad (6.27\text{a})
\end{aligned}$$

Note that problem (6.27) can be decomposed into two sets of subproblems, i.e., optimizing $\boldsymbol{\delta}$ and jointly optimizing $\boldsymbol{A}$ and $\boldsymbol{\tau}$. We first consider the following optimization problem for optimizing $\boldsymbol{A}$ and $\boldsymbol{\tau}$,

$$\begin{aligned}
\underset{\boldsymbol{A},\boldsymbol{\tau}}{\text{minimize}} \quad & \sum_{n=1}^{N}\sum_{k=1}^{K}\mu_k[n]\Big(\sum_{k=1}^{K}\tau_k[n]+a_k[n]t_k^{\text{comp}}\Big) \\
& +\sum_{k=1}^{K}\lambda_k\sum_{n=1}^{N}\Big(\frac{\tau_k[n]\sigma^2}{h_k[n]}\big(2^{\frac{a_k[n]s}{B\tau_k[n]}}-1\big)+a_k[n]E_k^{\text{comp}}[n]\Big) \\
& -\xi\sum_{n=1}^{N}\sum_{k=1}^{K}a_k[n]D_k^2 \\
\text{subject to} \quad & \text{Constraints (6.24)}, \qquad (6.28)
\end{aligned}$$

Problem (6.28) is convex with respect to $\{a_k[n]\}$ and $\{\tau_k[n]\}$, so the optimal solution is the one that satisfies the Karush-Kuhn-Tucker (KKT) conditions. To obtain the optimal $\boldsymbol{\tau}$ in problem (6.28), we have the following theorem.

Theorem 6.2 *By setting the first derivative of the objective function of problem* (6.28) *with respect to $\tau_k[n]$ to zeros, the optimal time allocation can be written as*

$$\tau_k^\star[n]=a_k[n]\left[\frac{s\ln 2}{B\big(1+\mathcal{W}\big(\frac{h_k[n]\sum_{k=1}^{K}\mu_k[n]}{\lambda_k\sigma^2 e}-\frac{1}{e}\big)}\right]^+,\forall k,\forall n, \qquad (6.29)$$

where $\mathcal{W}$ is the Lambert W function with $\mathcal{W}(x)e^{\mathcal{W}(x)}=x$ and $[x]^+\triangleq\max\{x,0\}$.

By substituting the obtained $\{\tilde{\tau}_k[n]\}$ into problem (6.28), we have

$$\begin{aligned} \underset{\mathbf{A}}{\text{minimize}} \ & \sum_{k=1}^{K}\sum_{n=1}^{N} g(\tilde{\tau}_k[n])a_k[n] \\ \text{subject to} \ & 0 \leq a_k[n] \leq 1, \forall k \in \mathcal{K}, \forall n \in \mathcal{N}, \end{aligned} \tag{6.30a}$$

where $\tilde{\tau}_k[n] = \frac{\tau_k^{\star}[n]}{a_k[n]}$, $g(\tilde{\tau}_k[n]) = \lambda_k\left(\frac{\tilde{\tau}_k[n]\sigma^2}{h_k[n]}\left(2^{\frac{s}{B\tilde{\tau}_k[n]}} - 1\right) + E_k^{\text{comp}}[n]\right) + \tilde{\tau}_k[n]\sum_{k=1}^{K} \mu_k[n] + \mu_k[n]t_k^{\text{comp}} - \xi D_k^2$. Problem (6.30) is a convex linear programming. It is evident that, to minimize the objective function in (6.30), the $a_k[n]$ corresponding to $g(\tilde{\tau}_k[n]) \leq 0$ should be set to 1. Therefore, the optimal solution for $\{a_k[n]\}$ is thus given by the following theorem.

Theorem 6.3 *For problem* (6.30), *the optimal device scheduling variables* **A** *can be expressed as*

$$a_k^{\star}[n] = \begin{cases} 1, \text{ if } g(\tilde{\tau}_k[n]) \leq 0, \\ 0, \text{ otherwise.} \end{cases} \tag{6.31}$$

As shown in the above theorem, those with substantial computational capacity, strong channel gain, and an extensive data-set should be given preference.

For the subproblem of optimizing $\boldsymbol{\delta}$, a linear programming, it is can rewritten as

$$\begin{aligned} \underset{\boldsymbol{\delta}}{\text{minimize}} \ & \sum_{n=1}^{N}\left(1 - \sum_{k=1}^{K}\mu_k[n]\right)\delta[n] \\ \text{subject to} \ & \delta[n] \geq (\|\boldsymbol{q}[n] - \boldsymbol{q}[n-1]\|_2)/V_{\max}, \forall n. \end{aligned} \tag{6.32}$$

For problem (6.32), since $\sum_{k=1}^{K}\mu_k[n] \leq 1$ holds, the optimal time slot variables are given by

$$\delta^{\star}[n] = \begin{cases} (\|\boldsymbol{q}[n] - \boldsymbol{q}[n-1]\|_2)/V_{\max}, \text{ if } \sum_{k=1}^{K}\mu_k[n] < 1, \\ b, \text{ if } \sum_{k=1}^{K}\mu_k[n] = 1, \end{cases} \tag{6.33}$$

where b can be any arbitrary real number which is not smaller than $\frac{\|\boldsymbol{q}[n]-\boldsymbol{q}[n-1]\|_2}{V_{\max}}$ since the objective function of problem (6.32) is not affected in this case. For simplicity, we set $b = \frac{\|\boldsymbol{q}[n]-\boldsymbol{q}[n-1]\|_2}{V_{\max}}$. In general, (6.33) cannot provide the optimal primal solution for problem (6.24) even with optimal dual variables. Nevertheless, with the above proposed solutions to problems (6.28) and (6.32), the dual function $g(\boldsymbol{\lambda}, \boldsymbol{\mu}, \xi)$ is obtained.

The dual problem (6.26) can be solved after obtaining $(\boldsymbol{\tau}^{\star}, \boldsymbol{\delta}^{\star}, \boldsymbol{A}^{\star})$. Despite the fact that the dual function $g(\boldsymbol{\lambda}, \boldsymbol{\mu}, \xi)$ is always convex by definition, it is generally non-differentiable. To address this challenge, a common subgradient-based method, such as projected subgradient descent method, can be applied to solve the (6.26).

Once $(\boldsymbol{\tau}^\star, \boldsymbol{\delta}^\star, \boldsymbol{A}^\star)$ is obtained, the dual problem (6.26) can be approached via the projected subgradient descent method. Specifically, the dual variables are updated in each iteration via

$$\mu_k^{\frac{1}{2}}[n] = [\mu_k[n] + \phi \sum_{k=1}^{K} \tau_k[n] + a_k[n] t_k^{\text{comp}} - \delta[n]]^+, \tag{6.34}$$

$$\mu_k[n] = \frac{\mu_k^{\frac{1}{2}}[n]}{\max\{1, \sum_{k=1}^{K} \mu_k^{\frac{1}{2}}[n]\}}, \tag{6.35}$$

$$\lambda_k = [\lambda_k + \phi \sum_{n=1}^{N} \Big(\frac{\tau_k[n]\sigma^2}{h_k[n]} \Big(2^{\frac{a_k[n]s}{B\tau_k[n]}} - 1 \Big) + a_k[n] E_k^{\text{comp}}[n] \Big) - E_k]^+, \tag{6.36}$$

$$\xi = [\xi + \phi(C - \sum_{n=0}^{N-1} \sum_{k=1}^{K} a_k[n+1] D_k^2)]^+, \tag{6.37}$$

where ϕ is a dynamically chosen step-size. Given $\boldsymbol{\lambda}^\star$, $\boldsymbol{\mu}^\star$, and $\xi^\star$, the optimal time allocation (i.e., $\boldsymbol{\tau}^\star$) and device scheduling variable (i.e., $\boldsymbol{A}^\star$) can be uniquely obtained by (6.29) and (6.31), respectively. By substituting $\boldsymbol{\tau}^\star$ and $\boldsymbol{A}^\star$ into the primal problem (6.24), we have

$$\begin{aligned} \underset{\boldsymbol{\delta}}{\text{minimize}} \quad & \sum_{n=1}^{N} \delta[n] \\ \text{subject to} \quad & \sum_{k=1}^{K} \tau_k[n] + a_k[n] t_k^{\text{comp}} \leq \delta[n], \forall k, \forall n \in \mathcal{N}, & \text{(6.38a)} \\ & \delta[n] \geq \frac{\|\boldsymbol{q}[n] - \boldsymbol{q}[n-1]\|_2}{V_{\max}}, \forall n. & \text{(6.38b)} \end{aligned}$$

Problem (6.38) is a linear programming and can be separated into N independent problems. The optimal time slot variables are given by

$$\delta^\star[n] = \max\left\{ \frac{\|\boldsymbol{q}[n] - \boldsymbol{q}[n-1]\|_2}{V_{\max}}, \sum_{k=1}^{K} \tau_k[n] + \max_k a_k[n] t_k^{\text{comp}} \right\}. \tag{6.39}$$

The details of solving the optimal solution of problem (6.24) can be summarized in Algorithm 7.

Algorithm 7: Dual method for problem (6.24)

1. **Input:** $K, N, \boldsymbol{Q}$
2. Initialize dual variables $\{\lambda_k = 1\}$, $\{\mu_k[n] = 1/K\}$
3. Repeat until $\boldsymbol{\lambda}$ and $\boldsymbol{\mu}$ converge within a prescribed accuracy:
   - Update the primal variables $\boldsymbol{A}$, $\boldsymbol{\tau}$, and $\boldsymbol{\delta}$ according to (6.31), (6.29), and (6.33).
   - Update the dual variables $\boldsymbol{\lambda}$, $\boldsymbol{\mu}$, ξ according to (6.35)–(6.37).
4. Update $\boldsymbol{\delta}^\star$ according to (6.39)
5. **Output:** $\boldsymbol{A}^\star$, $\boldsymbol{\tau}^\star$, and $\boldsymbol{\delta}^\star$

6.4.2 Trajectory Design

Given device scheduling and time allocation $\{\boldsymbol{A}, \boldsymbol{\delta}, \boldsymbol{\tau}\}$, the trajectory optimization subproblem is reduced to the following feasibility checking problem

$$\begin{aligned} \underset{\boldsymbol{Q}}{\text{find}} \quad & \boldsymbol{Q} \\ \text{subject to} \quad & \sum_{n=1}^{N} b_k[n] \|\boldsymbol{q}[n] - \mathbf{u}_k\|_2^2 \le \bar{E}_k, \forall k \in \mathcal{K}, && (6.40\text{a}) \\ & \text{Constraints (6.6), (6.7),} && (6.40\text{b}) \end{aligned}$$

where $\bar{E}_k \triangleq E_k - \sum_{n=1}^{N} a_k[n] E_k^{\text{comp}}[n] - \sum_{n=1}^{N} \frac{\tau_k[n]\sigma^2 H^2}{\beta_0}\big(2^{\frac{a_k[n]s}{B\tau_k[n]}} - 1\big)$ and $b_k[n] \triangleq \frac{\tau_k[n]\sigma^2}{\beta_0}\big(2^{\frac{a_k[n]s}{B\tau_k[n]}} - 1\big)$. According to (6.13), $\tau_k^\star[n]$ is a non-increasing function with respect to channel gain $h_k[n]$. Hence, to further reduce the time consumption while ensuring the feasibility of problem (6.40), we transform problem (6.40) into the following problem with an explicit objective function

$$\begin{aligned} \underset{\boldsymbol{Q}}{\text{minimize}} \quad & \sum_{n=1}^{N}\sum_{k=1}^{K} b_k[n] \|\boldsymbol{q}[n] - \mathbf{u}_k\|_2^2 \\ \text{subject to} \quad & \text{Constraints (6.6), (6.7), (6.40a),} && (6.41) \end{aligned}$$

which is a convex QCQP problem. It is evident that the feasible set of problem (6.41) is the same as that of problem (6.40). Hence, the solution to problem (6.41) is also feasible for problem (6.40).

The procedure for solving problem (6.41) is similar to that of solving problem (6.24). The details are omitted due to space limitations.

Table 6.1 Parameter settings for simulations

Number of devices	$K = 40$	Noise power	− 174 dBm/Hz
Channel gain at 1 m	$\beta_0 = -50$ dB	Uplink bandwidth	10 MHz
Communication rounds	$N = 4000$	Maximum distance	$\Delta q_{\max} = 15$ m

Algorithm 8: BCD-LD for Problem (6.19)

1. **Input:** M, N.
2. Initialize $\boldsymbol{Q}^0$. Let $r = 1$.
3. Repeat until the fractional decrease of the objective value is below a threshold ε:
   a. Solve problem (6.24) by applying Algorithm 7 for given $\boldsymbol{Q}^{r-1}$, and denote the optimal solution as $\{\boldsymbol{A}^r, \boldsymbol{\tau}^r, \boldsymbol{\delta}^r\}$.
   b. Solve problem (6.41) for given $\{\boldsymbol{A}^r, \boldsymbol{\tau}^r, \boldsymbol{\delta}^r\}$, and denote the optimal solution as $\boldsymbol{Q}^r$.
   c. Update $r = r + 1$.
4. **Output:** $\boldsymbol{A}$, $\boldsymbol{\tau}$, $\boldsymbol{\delta}$, and $\boldsymbol{Q}$.

6.4.3 Convergence and Complexity Analysis

In this section, building on the findings from the previous two subsections, we apply the BCD technique [15] to address problem (6.19). This is achieved by alternately solving Problem (6.24) and Problem (6.41) until convergence, as detailed in Algorithm 8. It is important to note that each block of variables is updated optimally in every iteration. The convergence of Algorithm 8 is demonstrated in the following proposition (Table 6.1).

Proposition 6.2 *With Algorithm 3, the objective value of problem* (6.19) *decreases as the number of iterations increases until convergence.*

6.5 Numerical Results

This section presents simulation results to demonstrate the effectiveness of the proposed algorithm for UAV-enabled FEEL system. The UAV's operational area is limited to a square region measuring [0, 400] m × [0, 400] m. It is assumed that the UAV flies at a constant altitude of $H = 100$ m, in compliance with regulations stating that commercial UAVs should not exceed an altitude of 400 feet (122 m) [11]. The initial position of the UAV is set at (200, 0, 100) m. The maximum speed of the UAV is $V_{\max} = 20$ m/s. Additionally, taking into account channel variations, K devices are divided into two clusters located in two circular areas: cluster A, consisting of $\lfloor 0.4K \rfloor$ devices and cluster B, comprising $K - \lfloor 0.4K \rfloor$ devices. Specifically, devices in cluster A and B are uniformly distributed within circles centered at (100, 100) m and (300, 300) m with a radius of 80m respectively.

The energy budget for devices in clusters A and B are denoted as E_A and E_B, respectively. All devices' computing chipsets have an effective capacitance coefficient $\alpha_1 = \ldots = \alpha_k = 10^{-28}$.

The multi-nominal logistic regression is leveraged to train the learning models, and the sample-wise loss function used for training is given by

$$f(\boldsymbol{w}; \boldsymbol{x}_{ki}, y_{ki}) = \sum_{c=1}^{C} \mathbb{I}_{\{y_{ki}=c\}} \log \frac{\exp(\boldsymbol{x}_{ki}\boldsymbol{w}_c^{\mathsf{T}})}{\sum_{j=1}^{C} \exp(\boldsymbol{x}_{ki}\boldsymbol{w}_j^{\mathsf{T}})}, \tag{6.42}$$

where C is the total number of label classes, $\boldsymbol{w} \triangleq [\boldsymbol{w}_1^{\mathsf{T}}, \ldots, \boldsymbol{w}_C^{\mathsf{T}}]^{\mathsf{T}}$, and $\mathbb{I}_{\{y_{ki}=c\}}$ is the indicator function. The partial gradient with respect to $\boldsymbol{w}_c$ is

$$\nabla_{\boldsymbol{w}_c} f(\boldsymbol{w}; \boldsymbol{x}_{ki}, y_{ki}) = -\left(\mathbb{I}_{\{y_{ki}=c\}} - \frac{\exp(\boldsymbol{x}_{ki}\boldsymbol{w}_c^{\mathsf{T}})}{\sum_{j=1}^{C} \exp(\boldsymbol{x}_{ki}\boldsymbol{w}_j^{\mathsf{T}})}\right)\boldsymbol{x}_{ki} \tag{6.43}$$

and the entire gradient is given by

$$\nabla_{\boldsymbol{w}} f(\boldsymbol{w}; \boldsymbol{x}_{ki}, y_{ki}) = \left[\nabla_{\boldsymbol{w}_1} f(\boldsymbol{w}; \boldsymbol{x}_{ki}, y_{ki}); \ldots; \nabla_{\boldsymbol{w}_C} f(\boldsymbol{w}; \boldsymbol{x}_{ki}, y_{ki})\right].$$

To assess the impact of various factors on the completion time of FEEL , we compare the performance of the proposed joint design with three benchmark schemes outlined below:

- **Full Scheduling**: The UAV-enabled FEEL training process uses full-scheduling transmission by setting $a_k[n] = 1, \forall k, \forall n$, while optimizing $\boldsymbol{\tau}, \boldsymbol{\delta}, \boldsymbol{Q}$. This approach acts as an upper performance bound, since it avoids introducing model aggregation errors.
- **Static UAV**: This scheme optimizes $\boldsymbol{A}, \boldsymbol{\tau}, \boldsymbol{\delta}$ and sets $\boldsymbol{q}[n] = \boldsymbol{q}_I, \forall n$. It underscores the significant communication straggler problem caused by a static UAV positioned far from all devices. Thus, it can be used to assess the effectiveness of UAV trajectory design in reducing the impact of stragglers on completion time cost.
- **Static UAV w/ HS**: This scheme uses a deterministic approach with fixed UAV location $\boldsymbol{Q}$ and device scheduling $\boldsymbol{A}$, setting $\boldsymbol{q}[n] = \boldsymbol{q}_I$ for all n, and selecting devices through a heuristic method that prioritizes those with higher channel gains as described in [12]. By satisfying constraint (6.19d) with equality, it only needs to optimize $\boldsymbol{\tau}, \boldsymbol{\delta}$ to minimize FEEL completion time. The scheme aims to evaluate the effectiveness of jointly designing UAV trajectory and device selection in mitigating the straggler effect on completion time and enhancing learning performance.

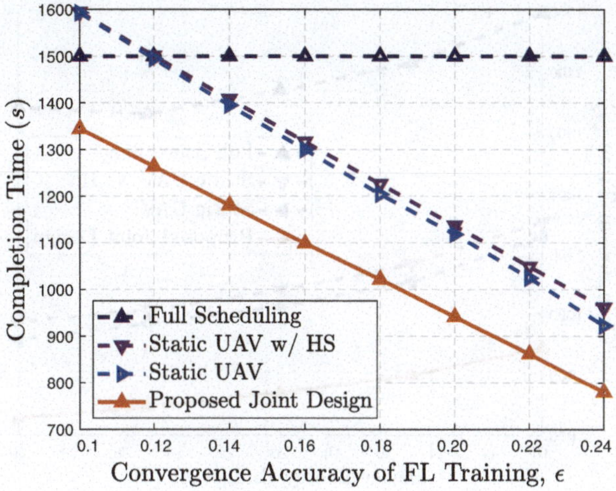

Fig. 6.3 Completion time versus convergence accuracy of FEEL

The initial UAV trajectory $\boldsymbol{Q}^0$ for BCD-LD and Full Scheduling is generated using the predetermined initial position, i.e., $\boldsymbol{q}^0[n] = \boldsymbol{q}_I, \forall n$, unless specified otherwise. Additionally, the global model parameters are initialized as a zero matrix.

6.5.1 Performance Comparison for Completion Time

Figure 6.3 illustrates the effect of the convergence accuracy of FEEL on the completion time when $E_k = 10$ J, $\forall k$. The performance of the full scheduling scheme remains unaffected by the convergence accuracy as it does not suffer from model aggregation loss. However, this scheme incurs excessively long completion times unless device scheduling is employed. For the other three schemes with device scheduling, the completion time decreases with ϵ. Among them, the Static UAV scheme outperforms the Static UAV w/ HS scheme because it mitigates aggregation error explosion by excluding weak devices. Nevertheless, when severe straggler issues arise, optimizing device scheduling alone brings only marginal performance gains. Fortunately, the proposed joint design scheme, which incorporates trajectory design and device scheduling, significantly reduces completion time compared to all reference benchmarks. This is achieved by leveraging the mobility of the UAV to minimize communication distances between scheduled devices and consistently achieve high transmission rates for local model uploading, thereby accelerating the model aggregation process. These results highlight the criticality of jointly designing trajectory and device scheduling in UAV-enabled systems to mitigate the adverse effects of communication stragglers on completion time.

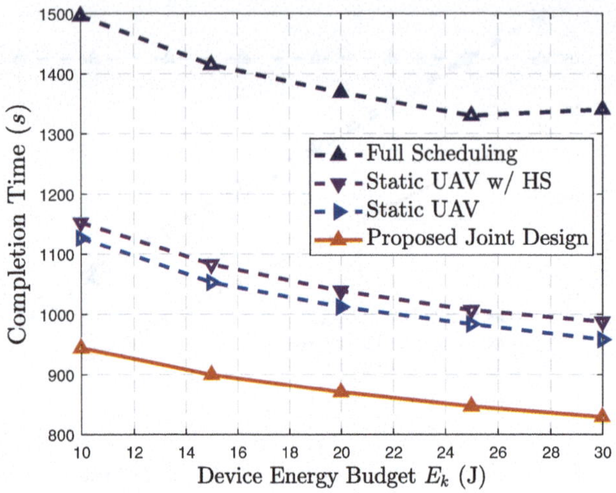

Fig. 6.4 Completion time versus device energy budget

Figure 6.4 illustrates the total mission completion time versus the device energy budget when $\epsilon = 0.2$. For all schemes, the total completion time decreases as E_k increases. The reason is that as E_k increases, the scheduled devices can afford more energy to upload local models with higher transmission rates. It is also observed that the proposed joint design achieves the smallest completion time compared to all benchmarks. This observation further demonstrates the superiority of the proposed scheme to mitigate the stragglers' effect.

6.5.2 Performance Comparison for FEEL

To demonstrate the performance of our proposed joint design for dealing with FEEL tasks, we train image classifier models on the widely-used CIFAR-10 dataset. Figure 6.5 shows the test accuracy versus convergence accuracy of FEEL training with different schemes when $E_k = 10$ J, $\forall k$. Figures 6.3 and 6.5 demonstrate that there is a fundamental trade-off between the mission completion time and the learning performance of FEEL. By exploiting the mobility of the UAV and flexibility of device scheduling, our proposed joint design significantly improves the test accuracy of FEEL compared to the benchmarks. However, by leveraging the mobility of the UAV and the flexibility of device scheduling, our proposed joint design significantly enhances the trade-off when compared to the benchmarks. Specifically, in our proposed joint design, the completion time is reduced by 38% compared to the full scheduling scheme, while the test accuracy only decreases by 0.71%.

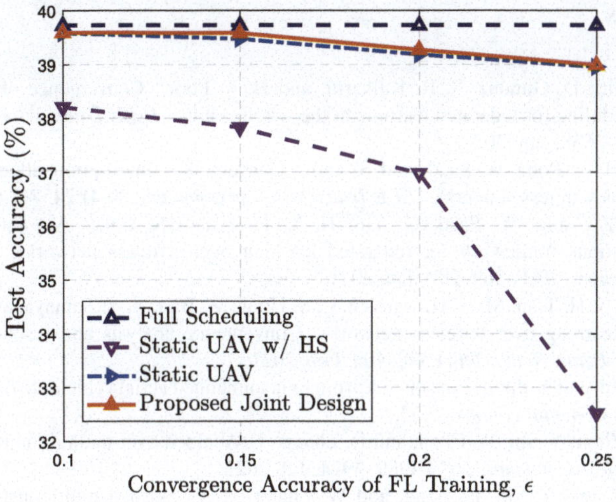

Fig. 6.5 Test accuracy versus convergence accuracy of FEEL

6.6 Conclusion

In this chapter, we studied the UAV-assisted FEEL system. The main objective of our research was to effectively address the straggler issue that occurs when the terrestrial PS is unavailable. To overcome this challenge, we leveraged the high altitude and mobility of a mobile UAV. In our approach, a mobile UAV served as a flying PS, facilitating the exchange of model parameters among distributed ground devices for training a shared FEEL model. Our focus was on jointly considering device scheduling, UAV trajectory, and time allocation in order to minimize the completion time required for FEEL to converge to the desired accuracy level. Before formulating the problem, we conducted a theoretical analysis to understand how device scheduling affects the convergence accuracy of FEEL training without assuming convexity. Additionally, we provided a convergence bound for the average norm of the global gradient. Building upon these convergence results, we formulated the problem of completion time minimization and proposed a joint design method to solve it. The simulation results demonstrated that our proposed joint design significantly reduced the completion time and enhanced the overall performance of FEEL.

References

1. M. M. Amiri, D. Gunduz, S. R. Kulkarni, and H. V. Poor. Convergence of update aware device scheduling for federated learning at the wireless edge. *IEEE Trans. Wireless Commun.*, 20(6):3643–3658, Jun. 2021.
2. M. Chen, H. V. Poor, W. Saad, and S. Cui. Convergence time optimization for federated learning over wireless networks. *IEEE Trans. Wireless Commun.*, 20(4):2457–2471, Apr. 2021.
3. M. Chen, Z. Yang, W. Saad, C. Yin, H. V. Poor, and S. Cui. A joint learning and communications framework for federated learning over wireless networks. *IEEE Trans. Wireless Comm.*, 20(1):269–283, Jan. 2021.
4. C. T. Dinh, N. H. Tran, M. N. H. Nguyen, C. S. Hong, W. Bao, A. Y. Zomaya, and V. Gramoli. Federated learning over wireless networks: Convergence analysis and resource allocation. *IEEE/ACM Trans. Netw.*, 29(1):398–409, Feb. 2021.
5. M. Fu, Y. Shi, and Y. Zhou. Federated learning via unmanned aerial vehicle. *IEEE Transactions on Wireless Communications*, 2023.
6. M. Fu, Y. Zhou, Y. Shi, W. Chen, and R. Zhang. UAV aided over-the-air computation. *IEEE Trans. Wireless Commun.*, 21(7):4909–4924, Jul. 2022.
7. M. Fu, Y. Zhou, Y. Shi, C. Jiang, and W. Zhang. UAV-assisted multi-cluster over-the-air computation. *IEEE Trans. Wireless Commun.*, 2022, doi: 10.1109/TWC.2022.3227768.
8. A. Imteaj, U. Thakker, S. Wang, J. Li, and M. H. Amini. A survey on federated learning for resource-constrained IoT devices. *IEEE Internet Things J.*, 9(1):1–24, Jan. 2022.
9. L. Li, L. Yang, X. Guo, Y. Shi, H. Wang, W. Chen, and K. B. Letaief. Delay analysis of wireless federated learning based on saddle point approximation and large deviation theory. *IEEE J. Select. Areas Commun.*, 39(12):3772–3789, Dec. 2021.
10. M. Li, N. Cheng, J. Gao, Y. Wang, L. Zhao, and X. Shen. Energy-efficient uav-assisted mobile edge computing: Resource allocation and trajectory optimization. *IEEE Transactions on Vehicular Technology*, 69(3):3424–3438, 2020.
11. M. Mozaffari, W. Saad, M. Bennis, Y.-H. Nam, and M. Debbah. A tutorial on UAVs for wireless networks: Applications, challenges, and open problems. *IEEE Commun. Surv. Tutorials*, 21(3):2334–2360, 2019.
12. J. Ren, Y. He, D. Wen, G. Yu, K. Huang, and D. Guo. Scheduling for cellular federated edge learning with importance and channel awareness. *IEEE Trans. Wireless Commun.*, 19(11):7690–7703, Nov. 2020.
13. S. Wang, S. Hosseinalipour, M. Gorlatova, C. G. Brinton, and M. Chiang. Uav-assisted online machine learning over multi-tiered networks: A hierarchical nested personalized federated learning approach. *IEEE Transactions on Network and Service Management*, 2022.
14. Q. Wu, Y. Zeng, and R. Zhang. Joint trajectory and communication design for multi-uav enabled wireless networks. *IEEE Transactions on Wireless Communications*, 17(3):2109–2121, 2018.
15. Y. Xu and W. Yin. A block coordinate descent method for regularized multiconvex optimization with applications to nonnegative tensor factorization and completion. *SIAM J. Imag. Sci.*, 6(3):1758–1789, 2013.
16. K. Yang, T. Jiang, Y. Shi, and Z. Ding. Federated learning via over-the-air computation. *IEEE Trans. Wireless Commun.*, 19(3):2022–2035, Mar. 2020.
17. C. Zhan and Y. Zeng. Completion time minimization for multi-uav-enabled data collection. *IEEE Transactions on Wireless Communications*, 18(10):4859–4872, 2019.
18. C. Zhan and Y. Zeng. Completion time minimization for multi-UAV-enabled data collection. *IEEE Trans. Wireless Commun.*, 18(10):4859–4872, Oct. 2019.
19. F. Zhou, Y. Wu, R. Q. Hu, and Y. Qian. Computation rate maximization in UAV-enabled wireless-powered mobile-edge computing systems. *IEEE J. Sel. Areas Commun.*, 36(9):1927–1941, Sept. 2018.
20. G. Zhu, Y. Wang, and K. Huang. Broadband analog aggregation for low-latency federated edge learning. *IEEE Trans. Wireless Commun.*, 19(1):491–506, Jan. 2020.

Chapter 7
Federated Edge Learning in Multi-Cell Wireless Networks

7.1 Overview

The existing research explores efficient strategies for collaborative learning and communication within a single-cell wireless network, specifically focusing on a single FEEL task [3, 9, 11, 14]. Given the rapid progress in connected intelligence, it is foreseeable that future wireless networks will operate with multiple FEEL tasks coexisting in multi-cell environments. Consequently, it becomes imperative to address the challenge of inter-cell interference, which significantly impacts the performance of such networks. Several recent studies have initiated preliminary discussions concerning the influence of inter-cell interference on device scheduling and resource allocation in wireless systems [8, 10, 18, 19]. However, it should be noted that these studies primarily concentrate on the assessment of inter-cell interference during uplink aggregation. In contrast, this chapter presents a departure from the existing literature by investigating the implications of inter-cell interference on both downlink and uplink model/gradient transmissions while considering multiple FEEL tasks across multi-cell wireless networks. This extension enables us to gain valuable insights into the dynamics of collaborative learning in multi-cell wireless network.

In this chapter, we explore AirComp-assisted FEEL in a multi-cell wireless network [13]. In this setup, each cell is assigned a unique FEEL task, and the devices within each cell transmit their local gradients to the corresponding BS using AirComp. During the training process, both the downlink and uplink transmissions are distorted by the receiver noise, channel fading, and inter-cell interference. To achieve the best learning performance for all cells, it is necessary to effectively manage inter-cell interference during both the downlink and uplink model/gradient transmissions. Besides, the transmission processes in different cells are coupled and hence simply focusing on single-cell optimization may deteriorate the learning performance in other cells, which makes it challenging to coordinate the downlink and uplink model/gradient transmissions in different cells. Hence, we develop

Y. Zhou et al., *Federated Edge Learning*, Wireless Networks,
https://doi.org/10.1007/978-3-031-96649-1_7

a cooperative optimization framework to balance the learning performance of different FEEL tasks in a multi-cell wireless network, which enables us to separately minimize the sum of gaps generated by the transmission distortions.

7.2 System Model

7.2.1 Learning Model

Figure 7.1 shows a multi-cell wireless network including M single-antenna BSs. The goal of BS $m \in \mathcal{M} = \{1, 2, \ldots, M\}$ is to train an ML model by coordinating K_m single-antenna devices located in cell m. To be specific, device $k \in \mathcal{K}_m = \{\sum_{l=1}^{m-1} K_l + 1, \sum_{l=1}^{m-1} K_l + 2, \ldots, \sum_{l=1}^{m-1} K_l + K_m\}$ is associated with BS m and has local dataset $\mathcal{D}_k$, where $\mathcal{K}_m \cap \mathcal{K}_l = \emptyset, \forall l \in \mathcal{M} \setminus \{m\}$. The local loss function of device $k \in \mathcal{K}_m$ evaluated at model parameter $\boldsymbol{w}_m \in \mathbb{R}^{D_m}$ is given by

$$F_{m,k}(\boldsymbol{w}_m) = \frac{1}{|\mathcal{D}_k|} \sum_{\boldsymbol{\xi}_i \in \mathcal{D}_k} f_m(\boldsymbol{w}_m; \boldsymbol{\xi}_i) \tag{7.1}$$

where $\boldsymbol{\xi}_i$ is the i-th labeled data sample, and $f_m(\boldsymbol{w}_m; \boldsymbol{\xi}_i)$ denotes the loss function quantifying the prediction error of model parameter $\boldsymbol{w}_m$ on data sample $\boldsymbol{\xi}_i$. We assume all the local datasets in the same cell own the same size, i.e., $|\mathcal{D}_k| = |\mathcal{D}_j|$, $\forall k, j \in \mathcal{K}_m$. Hence, the global empirical risk function for the learning task in cell

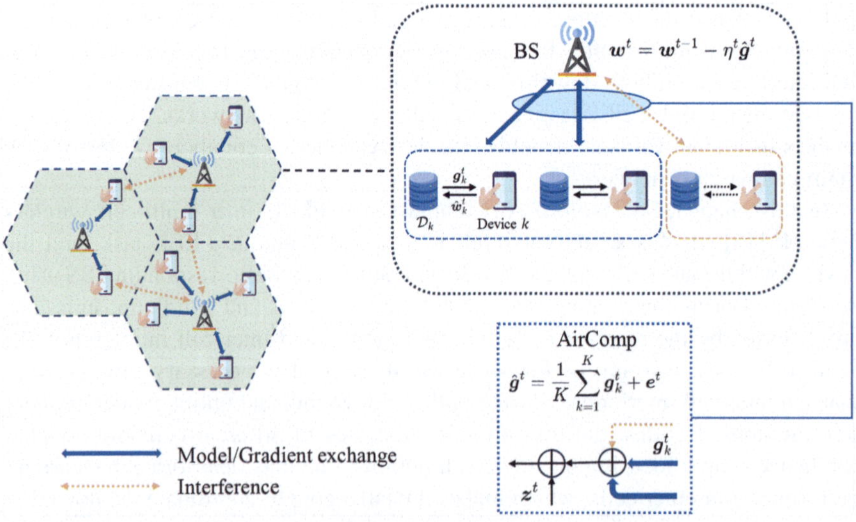

Fig. 7.1 Illustration of multiple FEEL tasks in a multi-cell wireless network

m can be expressed as

$$F_m(\boldsymbol{w}_m) = \frac{1}{K_m} \sum_{k \in \mathcal{K}_m} F_{m,k}(\boldsymbol{w}_m) \tag{7.2}$$

and the objective of FEEL is to obtain a global model such that

$$\boldsymbol{w}_m^\star = \arg \min_{\boldsymbol{w}_m \in \mathbb{R}^{D_m}} F_m(\boldsymbol{w}_m). \tag{7.3}$$

Within our multi-cell framework, each individual cell undertakes a distinct FEEL task.

7.2.2 *Communication Model*

Each cell iteratively performs the training steps to train a high-accuracy global model. Each training round consists of four stages: (1) *downlink model dissemination*, (2) *local gradient computation*, (3) *uplink gradient aggregation*, and (4) *global model update*, where stages (1) and (3) are carried out over wireless fading channels. Assume that the sample-wise loss functions and the dimensions of model parameters in different cells are the same, i.e., $f_m(\cdot) = f(\cdot)$ and $D_m = D$, $\forall m \in \mathcal{M}$. The details of the learning and communication design for the t-th round are given as follows.

7.2.2.1 Downlink Model Dissemination

BS m broadcasts its current global model $\boldsymbol{w}_m^{t-1}$ to the associated devices in set $\mathcal{K}_m$. Specifically, in the t-th round, BS m normalizes the current global model $\boldsymbol{w}_m^{t-1}$ in order to facilitate the transmit power control. This normalization process is defined as follows:

$$(\boldsymbol{s}_m^{\text{dl}})^t = \frac{\boldsymbol{w}_m^{t-1} - \bar{w}_m^{t-1}\mathbf{1}}{\nu_m^{t-1}} \tag{7.4}$$

Here, $\mathbf{1} = [1, 1, \ldots, 1]^\mathsf{T} \in \mathbb{R}^D$, and $\bar{w}_m^{t-1} \in \mathbb{R}$ and $\nu_m^{t-1} \in \mathbb{R}_{++}$ denote the mean and standard deviation of the D entries of the global model $\boldsymbol{w}_m^{t-1}$, respectively. These values are computed as:

$$\bar{w}_m^{t-1} = \frac{1}{D} \sum_{d=1}^{D} w_{m,d}^{t-1}, \; (\nu_m^{t-1})^2 = \frac{1}{D} \sum_{d=1}^{D} \left(w_{m,d}^{t-1} - \bar{w}_m^{t-1} \right)^2. \tag{7.5}$$

Hence, each element of the normalized model $(s_m^{\mathrm{dl}})^t$ has a zero mean and unit variance. This can be expressed as $\mathbb{E}[(s_{m,d}^{\mathrm{dl}})^t] = 0$ and $\mathbb{E}[((s_{m,d}^{\mathrm{dl}})^t)^2] = 1$, for all $d \in \{1, 2, \ldots, D\}$. For independent FEEL tasks in different cells, we can expect the cross-correlation between $(s_m^{\mathrm{dl}})^t$ and $(s_l^{\mathrm{dl}})^t$ to be $\mathbf{0}$, for all $m \neq l$.

Denote $(h_k^{\mathrm{dl}})^t \in \mathbb{C}$ and $(h_{l,k}^{\mathrm{dl}})^t \in \mathbb{C}$ as the downlink channel coefficients during the t-th round between device k and its home BS m, and between device k and its non-associated BS $l \in \mathcal{M} \setminus \{m\}$, respectively. These coefficients are complex numbers. We assume that each device can accurately estimate the channels connecting it to all the BSs using downlink pilot signaling. Additionally, each BS can acquire the local channel state information (CSI) of its associated devices through uplink feedback [4]. The channel gain for each link remains constant within one coherence time block, but it varies independently from one block to another. Each iteration consists of a downlink transmission block and an uplink transmission block. Each block has the capacity to transmit an entire model/gradient parameter. The communication protocol presented below primarily focuses on transmitting one representative dimension of the parameters. However, it can be extended to scenarios requiring multiple blocks to transmit high-dimensional model/gradient parameters. It's important to note that when the dimension of the model/gradient parameter is large, compression techniques can be used to enable transmission within one coherence block [1, 2].

The normalized global model, $(s_m^{\mathrm{dl}})^t$, is broadcasted by BS m to the devices in set $\mathcal{K}_m$. The signal received at device $k \in \mathcal{K}_m$ can be expressed as

$$\begin{aligned}(y_{k,d}^{\mathrm{dl}})^t &= (h_k^{\mathrm{dl}})^t \sqrt{(p_m^{\mathrm{dl}})^t} (s_{m,d}^{\mathrm{dl}})^t \\ &\quad + \underbrace{\sum_{l \in \mathcal{M}\setminus\{m\}} (h_{l,k}^{\mathrm{dl}})^t \sqrt{(p_l^{\mathrm{dl}})^t} (s_{l,d}^{\mathrm{dl}})^t}_{\text{Inter-cell interference}} + (z_{k,d}^{\mathrm{dl}})^t \end{aligned} \tag{7.6}$$

where $(p_m^{\mathrm{dl}})^t \in \mathbb{R}_+$ denotes the downlink transmit power at BS m and $(z_{k,d}^{\mathrm{dl}})^t \sim \mathcal{CN}(0, (\sigma_k^{\mathrm{dl}})^2)$ denotes the additive receiver Gaussian noise at device k. By assuming that BS m shares small-sized scalars $\{\bar{w}_m^{t-1}, v_m^{t-1}\}$ and $(p_m^{\mathrm{dl}})^t$ with its associated devices in an error-free manner before broadcasting the global model, device k can de-normalize the received signal as follows

$$\begin{aligned}(r_{k,d}^{\mathrm{dl}})^t &= \frac{((h_k^{\mathrm{dl}})^t)^\dagger v_m^{t-1}}{|(h_k^{\mathrm{dl}})^t|^2 \sqrt{(p_m^{\mathrm{dl}})^t}} (y_{k,d}^{\mathrm{dl}})^t + \bar{w}_m^{t-1} \\ &= w_{m,d}^{t-1} + \underbrace{\frac{((h_k^{\mathrm{dl}})^t)^\dagger v_m^{t-1}}{|(h_k^{\mathrm{dl}})^t|^2 \sqrt{(p_m^{\mathrm{dl}})^t}}}\end{aligned}$$

$$\times \underbrace{\left(\sum_{l \in \mathcal{M} \setminus \{m\}} (h_{l,k}^{\text{dl}})^t \sqrt{(p_l^{\text{dl}})^t} (s_{l,d}^{\text{dl}})^t + (z_{k,d}^{\text{dl}})^t \right)}_{(e_{k,d}^{\text{dl}})^t} \tag{7.7}$$

where $(e_{k,d}^{\text{dl}})^t$ is the downlink error induced by the receiver noise, channel fading, and inter-cell interference. The first term of $(e_{k,d}^{\text{dl}})^t$ is the error caused by inter-cell interference and the second term is the error induced by the additive noise. Therefore, the received model parameter at device k is

$$\widehat{\boldsymbol{w}}_k^t = \Re\left\{(\boldsymbol{r}_k^{\text{dl}})^t\right\} = \boldsymbol{w}_m^{t-1} + \Re\left\{(\boldsymbol{e}_k^{\text{dl}})^t\right\} \tag{7.8}$$

where $(\boldsymbol{r}_k^{\text{dl}})^t = [(r_{k,1}^{\text{dl}})^t, (r_{k,2}^{\text{dl}})^t, \ldots, (r_{k,D}^{\text{dl}})^t]^{\mathsf{T}}$ and $(\boldsymbol{e}_k^{\text{dl}})^t = [(e_{k,1}^{\text{dl}})^t, (e_{k,2}^{\text{dl}})^t, \ldots, (e_{k,D}^{\text{dl}})^t]^{\mathsf{T}}$.

7.2.2.2 Local Gradient Computation

After receiving estimated global model $\widehat{\boldsymbol{w}}_k^t$, device k computes local gradient $\boldsymbol{g}_k^t \in \mathbb{R}^D$ based on its local dataset $\mathcal{D}_k$ as

$$\boldsymbol{g}_k^t = \nabla F_k(\widehat{\boldsymbol{w}}_k^t) = \frac{1}{|\mathcal{D}_k|} \sum_{\xi_i \in \mathcal{D}_k} \nabla f(\widehat{\boldsymbol{w}}_k^t; \xi_i). \tag{7.9}$$

By uniformly sampling a batch of data $(\mathcal{D}_k^{\text{bs}})^t \subset \mathcal{D}_k$, the mini-batch stochastic gradient can be written as

$$\widetilde{\boldsymbol{g}}_k^t = \frac{1}{|(\mathcal{D}_k^{\text{bs}})^t|} \sum_{\xi_i \in (\mathcal{D}_k^{\text{bs}})^t} \nabla f(\widehat{\boldsymbol{w}}_k^t; \xi_i) \tag{7.10}$$

where $\mathbb{E}_{(\mathcal{D}_k^{\text{bs}})^t}\left[\widetilde{\boldsymbol{g}}_k^t\right] = \boldsymbol{g}_k^t$.

We mainly adopt (7.9) for the convergence analysis and device $k \in \mathcal{K}_m$ uploads its local gradient parameter $\boldsymbol{g}_k^t$ to its home BS m in the uplink transmission.

7.2.2.3 Uplink Gradient Aggregation

For global model update, BS m aims to obtain an arithmetic mean of the local gradients of all devices in set $\mathcal{K}_m$

$$\boldsymbol{g}_m^t = \frac{1}{K_m} \sum_{k \in \mathcal{K}_m} \boldsymbol{g}_k^t. \tag{7.11}$$

We utilize AirComp for uplink gradient aggregation in order to improve communication efficiency. With AirComp, BS m is able to directly obtain a noisy version of the arithmetic mean. Similar to downlink transmission, device k first normalizes its local gradient as follows:

$$(\boldsymbol{s}_k^{\mathrm{ul}})^t = \frac{\boldsymbol{g}_k^t - \bar{g}_k^t \mathbf{1}}{\upsilon_k^t}, \tag{7.12}$$

where $\bar{g}_k^t \in \mathbb{R}$ and $\upsilon_k^t \in \mathbb{R}_{++}$ denote the mean and standard deviation of D entries of local gradient $\boldsymbol{g}_k^t$, respectively. They are defined as

$$\bar{g}_k^t = \frac{1}{D}\sum_{d=1}^{D} g_{k,d}^t,\ (\upsilon_k^t)^2 = \frac{1}{D}\sum_{d=1}^{D}\left(g_{k,d}^t - \bar{g}_k^t\right)^2. \tag{7.13}$$

As a result, the transmitted signals satisfy $\mathbb{E}[(s_{k,d}^{\mathrm{ul}})^t] = 0$ and $\mathbb{E}[((s_{k,d}^{\mathrm{ul}})^t)^2] = 1$, $\forall d \in \{1, 2, \ldots, D\}$, $\forall k \in \mathcal{K}_m$. Assuming that the local gradients of different devices are independent, we have $\mathbb{E}[(\boldsymbol{s}_k^{\mathrm{ul}})^t((\boldsymbol{s}_{k'}^{\mathrm{ul}})^t)^{\mathsf{T}}] = \mathbf{0}$, $\forall k \neq k'$.

Let $(h_k^{\mathrm{ul}})^t \in \mathbb{C}$ and $(h_{k,l}^{\mathrm{ul}})^t \in \mathbb{C}$, $\forall k \in \mathcal{K}_m$, denote the uplink channel coefficients during the t-th round between device k and its home BS m, and between device k and its non-associated BS $l \in \mathcal{M} \setminus \{m\}$, respectively. By the concurrent transmissions, the signal received at BS m can be written as

$$(y_{m,d}^{\mathrm{ul}})^t = \sum_{k\in\mathcal{K}_m} (h_k^{\mathrm{ul}})^t b_k^t (s_{k,d}^{\mathrm{ul}})^t + (z_{m,d}^{\mathrm{ul}})^t + \underbrace{\sum_{l\in\mathcal{M}\setminus\{m\}} \sum_{k'\in\mathcal{K}_l} (h_{k',m}^{\mathrm{ul}})^t b_{k'}^t (s_{k',d}^{\mathrm{ul}})^t}_{\text{Inter-cell interference}} \tag{7.14}$$

where $b_k^t \in \mathbb{C}$ is the transmit scalar at device k and $(z_{m,d}^{\mathrm{ul}})^t \sim \mathcal{CN}(0, (\sigma_m^{\mathrm{ul}})^2)$ is the additive receiver Gaussian noise at BS m. In order to compensate for the phase distortion caused by complex channel responses, the transmit scalar is determined as $b_k^t = \frac{((h_k^{\mathrm{ul}})^t)^\dagger}{|(h_k^{\mathrm{ul}})^t|}\sqrt{(p_k^{\mathrm{ul}})^t}$, where $(p_k^{\mathrm{ul}})^t \in \mathbb{R}_+$ represents the uplink transmit power at device k. As a result, (7.14) can be simplified to

$$(y_{m,d}^{\mathrm{ul}})^t = \sum_{k\in\mathcal{K}_m} |(h_k^{\mathrm{ul}})^t| \sqrt{(p_k^{\mathrm{ul}})}(s_{k,d}^{ul}) + (z_{m,d}^{ul}) + \tag{7.15}$$

$$* + \sum_{l\in M/m} \sum_{k' in K_l} \frac{(h_{k',m}^{ul}))((h_{k'})^t dag}{|(h_{k'})^t|} sqrt(p_{k'})^t (s_{k',d}^{ul}) \tag{7.16}$$

By assuming that the devices report scalars $\{\bar{g}_k^t, \upsilon_k^t\}_{k\in\mathcal{K}_m}$ to their home BSs in an error-free manner, BS m can obtain the scaled signal

$$\begin{aligned}(r_{m,d}^{\text{ul}})^t &= \frac{1}{K_m}\left(\frac{(y_{m,d}^{\text{ul}})^t}{\sqrt{c_m^t}} + \sum_{k\in\mathcal{K}_m} \bar{g}_k^t\right)\\ &= \frac{1}{K_m}\sum_{k\in\mathcal{K}_m} g_{k,d}^t + \frac{1}{K_m}\\ &\quad \times \underbrace{\left(\sum_{k\in\mathcal{K}_m}\left(\frac{|(h_k^{\text{ul}})^t|\sqrt{(p_k^{\text{ul}})^t}}{\sqrt{c_m^t}} - \upsilon_k^t\right)(s_{k,d}^{\text{ul}})^t + \frac{(z_{m,d}^{\text{ul}})^t}{\sqrt{c_m^t}}\right.}\\ &\quad \underbrace{\left. + \sum_{l\in\mathcal{M}\backslash\{m\}}\sum_{k'\in\mathcal{K}_l}\frac{(h_{k',m}^{\text{ul}})^t((h_{k'}^{\text{ul}})^t)^{\dagger}\sqrt{(p_{k'}^{\text{ul}})^t}(s_{k',d}^{\text{ul}})^t}{|(h_{k'}^{\text{ul}})^t|\sqrt{c_m^t}}\right)}_{(e_{m,d}^{\text{ul}})^t} \end{aligned} \tag{7.17}$$

where $c_m^t \in \mathbb{R}_{++}$ is the receive normalizing factor for signal power alignment and noise power suppression, and $(e_{m,d}^{\text{ul}})^t$ is the uplink aggregation error caused by the receiver noise, channel fading, and inter-cell interference. The initial component of $(e_{m,d}^{\text{ul}})^t$ symbolizes the discrepancy arising from variations in channel fading and power control. The subsequent term signifies the error linked to the presence of noise at the receiver, while the third element characterizes the error associated with inter-cell interference. Therefore, the average of the local gradients received at BS m is given by

$$\widehat{\boldsymbol{g}}_m^t = \Re\left\{(\boldsymbol{r}_m^{\text{ul}})^t\right\} = \frac{1}{K_m}\left(\sum_{k\in\mathcal{K}_m} \boldsymbol{g}_k^t + \Re\left\{(\boldsymbol{e}_m^{\text{ul}})^t\right\}\right) \tag{7.18}$$

where $(\boldsymbol{r}_m^{\text{ul}})^t = [(r_{m,1}^{\text{ul}})^t, (r_{m,2}^{\text{ul}})^t, \ldots, (r_{m,D}^{\text{ul}})^t]^{\mathsf{T}}$ and $(\boldsymbol{e}_m^{\text{ul}})^t = [(e_{m,1}^{\text{ul}})^t, (e_{m,2}^{\text{ul}})^t, \ldots, (e_{m,D}^{\text{ul}})^t]^{\mathsf{T}}$.

7.2.2.4 Global Model Update

Based on the average of local gradients received at BS m given in (7.18), the global model maintained by BS m can be updated as

$$\boldsymbol{w}_m^t = \boldsymbol{w}_m^{t-1} - \eta_m^t \widehat{\boldsymbol{g}}_m^t \tag{7.19}$$

where η_m^t denotes the learning rate of cell m.

7.3 Convergence Analysis and Problem Formulation

This section presents the convergence analysis of AirComp-assisted FEEL in multi-cell wireless networks, followed by constructing a cooperative optimization problem that aims at achieving desirable learning performance across various FEEL tasks.

7.3.1 Convergence Analysis

We first make the following two standard assumptions.

Assumption 15 *Global loss function $F_m(\cdot)$, $\forall m \in \mathcal{M}$, is lower bounded, i.e., $F_m(\boldsymbol{w}) \geq F_m(\boldsymbol{w}^\star) > -\infty$, $\forall \boldsymbol{w} \in \mathbb{R}^D$.*

Assumption 16 *$\nabla F_{m,k}(\cdot)$, $\forall k \in \mathcal{K}_m$, $\forall m \in \mathcal{M}$ are Lipschitz continuous with constant $L > 0$, i.e., for any $\boldsymbol{w}$, $\widetilde{\boldsymbol{w}} \in \mathbb{R}^D$,*

$$\left\|\nabla F_{m,k}(\boldsymbol{w}) - \nabla F_{m,k}(\widetilde{\boldsymbol{w}})\right\| \leq L \left\|\boldsymbol{w} - \widetilde{\boldsymbol{w}}\right\|, \tag{7.20}$$

which is also equivalent to

$$F_{m,k}(\boldsymbol{w}) - F_{m,k}(\widetilde{\boldsymbol{w}}) \leq \left\langle \nabla F_{m,k}(\widetilde{\boldsymbol{w}}), \boldsymbol{w} - \widetilde{\boldsymbol{w}} \right\rangle + \frac{L}{2} \left\|\boldsymbol{w} - \widetilde{\boldsymbol{w}}\right\|^2. \tag{7.21}$$

Based on the above assumptions, we present an upper bound of the time-average norm of the global gradients in the following theorem.

Theorem 7.1 *In cell m, by setting $0 < \eta_m^t \equiv \eta_m < \frac{1}{L}$, the time-average norm of the global gradients after T rounds is upper bounded as*

$$\begin{aligned} \frac{1}{T}\sum_{t=0}^{T-1} \mathbb{E}\left[\left\|\nabla F_m(\boldsymbol{w}_m^t)\right\|^2\right] \leq & \underbrace{\frac{2}{\eta_m T}\left(F_m(\boldsymbol{w}_m^0) - F_m(\boldsymbol{w}_m^\star)\right)}_{\textit{Initial gap}} \\ & \underbrace{+ \frac{1}{T}\sum_{t=0}^{T-1}\sum_{d=1}^{D}\left(\frac{L^2}{K_m}\sum_{k\in\mathcal{K}} \mathbb{E}\left[\Re\left\{(e_{k,d}^{\mathrm{dl}})^{t+1}\right\}^2\right]\right.} \\ & \underbrace{\left. + \frac{L\eta_m}{K_m^2}\mathbb{E}\left[\Re\left\{(e_{m,d}^{\mathrm{ul}})^{t+1}\right\}^2\right]\right)}_{\textit{Error-induced gap}}, \ \forall m \in \mathcal{M} \end{aligned} \tag{7.22}$$

where the expectation is taken over normalized transmit symbols and receiver noise.

According to Theorem 7.1, the upper bound of $\frac{1}{T}\sum_{t=0}^{T-1}\mathbb{E}\left[\|\nabla F_m(\boldsymbol{w}_m^t)\|^2\right]$ given in (7.22) consists of two parts, i.e., *initial gap* and *error-induced gap*. When T is large, the upper bound is dominated by the error-induced gap. Inspired by this observation, we try to minimize the error-induced gap in each time slot for transmitting one-dimension model/gradient parameter in each cell, given by

$$\frac{L^2}{K_m}\sum_{k\in\mathcal{K}_m}\mathbb{E}\left[\Re\left\{(e_{k,d}^{\mathrm{dl}})^t\right\}^2\right]+\frac{L\eta_m}{K_m^2}\mathbb{E}\left[\Re\left\{(e_{m,d}^{\mathrm{ul}})^t\right\}^2\right],$$
$$\forall d\in\{1,2,\ldots,D\},\ \forall t\in\mathcal{T},\ \forall m\in\mathcal{M}. \tag{7.23}$$

The reason is that the error-induced gap is minimized when (7.23) is minimized. However, solely prioritizing the reduction of (7.23) in individual cells could result in significant interference between cells within multi-cell wireless networks, thereby compromising the learning capabilities of other cells. As a result, it becomes essential to adopt a collaborative approach that ensures a harmonious balance of learning performance across multiple cells while carrying out federated learning tasks.

By examining the second component in (7.23), it becomes evident that leveraging gradient aggregation during the uplink transmission can yield advantageous outcomes when employing a low learning rate η_m. This approach effectively mitigates the influence of uplink aggregation errors. The reason is that the uplink aggregation error is incorporated into the gradients depicted in (7.18), allowing for the learning rate η_m to reconfigure the error within (7.19).

7.3.2 Problem Formulation

For presentation clarity, we omit the time indices in the following problem formulation. In order to achieve equilibrium in learning performance across various federated learning tasks, we introduce the concept of the disparity region, denoted as $\mathcal{D}$. This region comprises tuples $(\Delta_1,\Delta_2,\ldots,\Delta_M)$ that signify the concurrent occurrence of error-induced gaps in all cells, subject to constraints imposed by restricted wireless communication resources. Considering that the model/gradient parameters are transmitted via the analog modulation scheme on a shared radio channel, communication errors predominantly emerge from signal misalignment and distortion arising from limitations in transmit power budgets. Specifically, the gap region, $\mathcal{G}$, can be expressed as

$$\mathcal{G}=\bigcup\{(\Delta_1,\Delta_2,\ldots,\Delta_M)\mid\Delta_m\geq\mathsf{Gap}_m,\ \forall m\in\mathcal{M}\} \tag{7.24}$$

where

$$\mathsf{Gap}_m = \underbrace{\frac{L^2 E_m^{\mathrm{dl}}}{K_m}}_{\mathsf{Gap}_m^{\mathrm{dl}}} + \underbrace{\frac{L\eta_m E_m^{\mathrm{ul}}}{K_m^2}}_{\mathsf{Gap}_m^{\mathrm{ul}}} \tag{7.25}$$

denotes the error-induced gap with

$$\begin{aligned} E_m^{\mathrm{dl}} &= \sum_{k\in\mathcal{K}_m} \mathbb{E}\left[\Re\left\{e_{k,d}^{\mathrm{dl}}\right\}^2\right] \\ &= \sum_{k\in\mathcal{K}_m} v_m^2 \left(\sum_{l\in\mathcal{M}\setminus\{m\}} \frac{\Re\left\{(h_k^{\mathrm{dl}})^\dagger h_{l,k}^{\mathrm{dl}}\right\}^2 p_l^{\mathrm{dl}}}{|h_k^{\mathrm{dl}}|^4 p_m^{\mathrm{dl}}} + \frac{(\sigma_k^{\mathrm{dl}})^2}{2|h_k^{\mathrm{dl}}|^2 p_m^{\mathrm{dl}}} \right) \end{aligned} \tag{7.26}$$

and

$$\begin{aligned} E_m^{\mathrm{ul}} = \mathbb{E}\left[\Re\left\{e_{m,d}^{\mathrm{ul}}\right\}^2\right] &= \sum_{k\in\mathcal{K}_m} \left(\frac{|h_k^{\mathrm{ul}}|\sqrt{p_k^{\mathrm{ul}}}}{\sqrt{c_m}} - \upsilon_k \right)^2 \\ &+ \sum_{l\in\mathcal{M}\setminus\{m\}} \sum_{k'\in\mathcal{K}_l} \frac{\Re\left\{h_{k',m}^{\mathrm{ul}}(h_{k'}^{\mathrm{ul}})^\dagger\right\}^2 p_{k'}^{\mathrm{ul}}}{|h_{k'}^{\mathrm{ul}}|^2 c_m} + \frac{(\sigma_m^{\mathrm{ul}})^2}{2c_m}. \end{aligned} \tag{7.27}$$

Here, (7.26) illustrates the effect of the sum of downlink dissemination errors at the devices in cell m and (7.27) illustrates the impact of the uplink aggregation error at BS m on Gap_m. Subsequently, Gap_m is determined by

$$\begin{aligned} \mathcal{V} = \Bigg\{ \Bigg(\bigcup_{m\in\mathcal{M}} \left\{ p_m^{\mathrm{dl}}, c_m, \left\{p_k^{\mathrm{ul}}\right\}_{k\in\mathcal{K}_m} \right\} \Bigg) \Bigg| \, 0 \le p_m^{\mathrm{dl}} \le P_m^{\mathrm{dl}}, \\ c_m > 0, 0 \le p_k^{\mathrm{ul}} \le P_k^{\mathrm{ul}}, \forall k \in \mathcal{K}_m, \forall m \in \mathcal{M} \Bigg\} \end{aligned} \tag{7.28}$$

where P_m^{dl} and P_k^{ul} is the maximum transmit power budgets in the downlink and uplink, respectively.

Since the value of Gap_m in each cell is subject to the influence of inter-cell interference, there exists a trade-off in performance among different federated learning tasks conducted across multiple cells. Specifically, the reduction of error-induced gaps in one cell may result in an increase in the error-induced gaps of other cells. Hence, our objective is to identify a feasible set denoted as $\mathcal{V}$ that allows us to approach the Pareto boundary $\mathcal{P}$ within the gap region $\mathcal{G}$ in order to

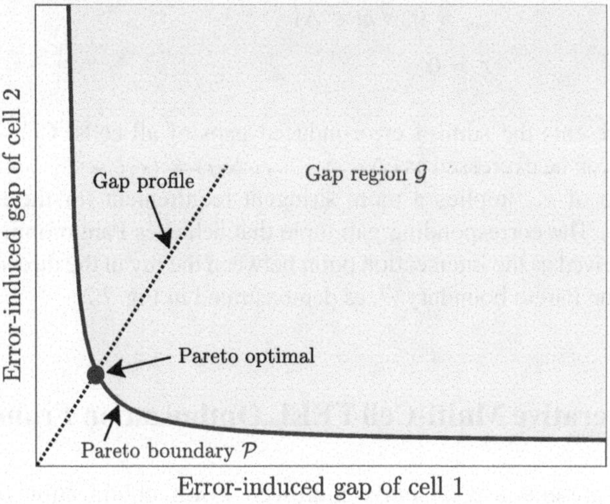

Fig. 7.2 Illustration of Pareto boundary $\mathcal{P}$ of gap region $\mathcal{G}$ in a two-cell wireless network

achieve a balance in learning performance across multiple cells. The concept of Pareto optimality for a tuple $(\Delta_1, \Delta_2, \ldots, \Delta_M)$ is defined as follows [6].

Definition 7.1 Tuple $(\Delta_1, \Delta_2, \ldots, \Delta_M)$ is Pareto optimal if there is no other tuple $(\widetilde{\Delta}_1, \widetilde{\Delta}_2, \ldots, \widetilde{\Delta}_M)$ with $(\widetilde{\Delta}_1, \widetilde{\Delta}_2, \ldots, \widetilde{\Delta}_M) \preceq (\Delta_1, \Delta_2, \ldots, \Delta_M)$ and $(\widetilde{\Delta}_1, \widetilde{\Delta}_2, \ldots, \widetilde{\Delta}_M) \neq (\Delta_1, \Delta_2, \ldots, \Delta_M)$, where $\preceq$ denotes the component-wise inequality.

An illustrative example featuring two cells is depicted in Fig. 7.2, where the shaded region represents the gap region $\mathcal{G}$, with its lower-left boundary representing the Pareto boundary $\mathcal{P}$. On this boundary, reducing the error-induced gap of one cell comes at the expense of increasing the error-induced gap of the other cell. To characterize the Pareto boundary, we employ the profiling technique [4, 16], which involves coordinating all base stations (BSs) to minimize the total sum of error-induced gaps across all cells. In particular, we introduce a profiling vector $\boldsymbol{\kappa} = [\kappa_1, \kappa_2, \ldots, \kappa_M]$, satisfying $\kappa_m \geq 0$ for all $m \in \mathcal{M}$ and $\sum_{m \in \mathcal{M}} \kappa_m = 1$. By solving the following problem, we can obtain the tuple of gaps residing on the Pareto boundary $\mathcal{P}$.

$$\underset{\substack{\zeta, \{c_m\}, \\ \{p_m^{\text{dl}}\}, \{p_k^{\text{ul}}\}}}{\text{minimize}} \quad \zeta \tag{7.29a}$$

$$\text{subject to} \quad \text{Gap}_m \leq \kappa_m \zeta, \ \forall m \in \mathcal{M} \tag{7.29b}$$

$$0 \leq p_m^{\text{dl}} \leq P_m^{\text{dl}}, \ \forall m \in \mathcal{M} \tag{7.29c}$$

$$0 \leq p_k^{\text{ul}} \leq P_k^{\text{ul}}, \ \forall k \in \mathcal{K}_m, \ \forall m \in \mathcal{M} \tag{7.29d}$$

$$c_m > 0, \forall m \in \mathcal{M} \tag{7.29e}$$

$$\zeta \geq 0 \tag{7.29f}$$

where ζ represents the sum of error-induced gaps of all cells. Consequently, the tuple of gaps can be expressed as $(\Delta_1, \Delta_2, \ldots, \Delta_M) = (\kappa_1\zeta, \kappa_2\zeta, \ldots, \kappa_M\zeta)$, where a lower value of κ_m implies a more stringent requirement for the error-induced gap of cell m. The corresponding gap tuple that achieves Pareto optimality can be visually perceived as the intersection point between the ray in the direction of $\boldsymbol{\kappa}$ (gap profile) and the Pareto boundary $\mathcal{P}$, as demonstrated in Fig. 7.2.

7.4 Cooperative Multi-Cell FEEL Optimization Framework

This section presents a cooperative multi-cell FEEL optimization framework to balance the learning performance among different FEEL tasks in multiple cells.

Let $\zeta = \zeta_0^{\mathrm{dl}} + \zeta_0^{\mathrm{ul}}$. We utilize ζ_0^{dl} and ζ_0^{ul} to quantify the sum of instantaneous error-induced gaps generated by downlink and uplink transmissions, respectively. We then rewrite problem (7.29) as

$$\underset{\zeta_0^{\mathrm{dl}}, \zeta_0^{\mathrm{ul}}, \{c_m\}, \{p_m^{\mathrm{dl}}\}, \{p_k^{\mathrm{ul}}\}}{\text{minimize}} \quad \zeta_0^{\mathrm{dl}} + \zeta_0^{\mathrm{ul}} \tag{7.30a}$$

$$\text{subject to} \quad \mathsf{Gap}_m^{\mathrm{dl}} \leq \kappa_m \zeta_0^{\mathrm{dl}}, \ \forall m \in \mathcal{M} \tag{7.30b}$$

$$\mathsf{Gap}_m^{\mathrm{ul}} \leq \kappa_m \zeta_0^{\mathrm{ul}}, \ \forall m \in \mathcal{M} \tag{7.30c}$$

$$\zeta_0^{\mathrm{dl}} \geq 0 \tag{7.30d}$$

$$\zeta_0^{\mathrm{ul}} \geq 0 \tag{7.30e}$$

$$\text{constraints (7.29c), (7.29d), (7.29e).} \tag{7.30f}$$

It is evident that the downlink and uplink transmissions are disjoint, which means that we can solve this problem separately.

7.4.1 *Cooperative Downlink Transmission*

The optimization problem regarding downlink transmission can be written as

$$\underset{\zeta_0^{\mathrm{dl}}, \{p_m^{\mathrm{dl}}\}}{\text{minimize}} \quad \zeta_0^{\mathrm{dl}} \tag{7.31a}$$

$$\text{subject to} \quad \text{constraints (7.29c), (7.30b), (7.30d).} \tag{7.31b}$$

Substituting (7.26) into (7.30b), we obtain

$$\underset{\zeta^{\mathrm{dl}},\{p_m^{\mathrm{dl}}\}}{\text{minimize}} \quad \zeta^{\mathrm{dl}} \tag{7.32a}$$

$$\text{subject to} \quad \sum_{l\in\mathcal{M}\setminus\{m\}} p_l^{\mathrm{dl}} \sum_{k\in\mathcal{K}_m} \frac{\Re\left\{(h_k^{\mathrm{dl}})^\dagger h_{l,k}^{\mathrm{dl}}\right\}^2}{|h_k^{\mathrm{dl}}|^4} + \sum_{k\in\mathcal{K}_m} \frac{(\sigma_k^{\mathrm{dl}})^2}{2|h_k^{\mathrm{dl}}|^2} \le \frac{\kappa_m^{\mathrm{dl}}\zeta^{\mathrm{dl}}}{\nu_m^2} p_m^{\mathrm{dl}}, \ \forall m \in \mathcal{M} \tag{7.32b}$$

$$0 \le p_m^{\mathrm{dl}} \le P_m^{\mathrm{dl}}, \ \forall m \in \mathcal{M} \tag{7.32c}$$

$$\zeta^{\mathrm{dl}} \ge 0 \tag{7.32d}$$

where $\zeta^{\mathrm{dl}} = \frac{\zeta_0^{\mathrm{dl}}}{L^2}$ and $\kappa_m^{\mathrm{dl}} = \kappa_m K_m$. Problem (7.32) can be solved by tackling a sequence of feasibility detection problems, given ζ^{dl}. Specifically, for any given ζ^{dl}, problem (7.32) is reduced to

$$\text{find} \quad \{p_m^{\mathrm{dl}}\} \tag{7.33a}$$

$$\text{subject to} \quad \sum_{l\in\mathcal{M}\setminus\{m\}} p_l^{\mathrm{dl}} \sum_{k\in\mathcal{K}_m} \frac{\Re\left\{(h_k^{\mathrm{dl}})^\dagger h_{l,k}^{\mathrm{dl}}\right\}^2}{|h_k^{\mathrm{dl}}|^4} + \sum_{k\in\mathcal{K}_m} \frac{(\sigma_k^{\mathrm{dl}})^2}{2|h_k^{\mathrm{dl}}|^2} \le \frac{\kappa_m^{\mathrm{dl}}\zeta^{\mathrm{dl}}}{\nu_m^2} p_m^{\mathrm{dl}}, \ \forall m \in \mathcal{M} \tag{7.33b}$$

$$0 \le p_m^{\mathrm{dl}} \le P_m^{\mathrm{dl}}, \ \forall m \in \mathcal{M}. \tag{7.33c}$$

Here, constraints (7.33b) can be rewritten as

$$\sqrt{\varpi_{m,0}^2 + \sum_{l\in\mathcal{M}} \varpi_{m,l}^2 p_l^{\mathrm{dl}}} \le \sqrt{\frac{\kappa_m^{\mathrm{dl}}\zeta^{\mathrm{dl}}}{\nu_{0,m}^2} p_m^{\mathrm{dl}}}, \ \forall m \in \mathcal{M} \tag{7.34}$$

where

$$\varpi_{m,0}^2 = \sum_{k\in\mathcal{K}_m} \frac{(\sigma_k^{\mathrm{dl}})^2}{2|h_k^{\mathrm{dl}}|^2}, \ \varpi_{m,m}^2 = 0$$

$$\varpi_{m,l}^2 = \sum_{k\in\mathcal{K}_m} \frac{\Re\left\{(h_k^{\mathrm{dl}})^\dagger h_{l,k}^{\mathrm{dl}}\right\}^2}{|h_k^{\mathrm{dl}}|^4}, \ \forall l \in \mathcal{M} \setminus \{m\}. \tag{7.35}$$

Algorithm 9: Cooperative downlink optimization for problem (7.31)

Input: Standard deviations of global models $\{\upsilon_m\}$, profiling vector $\boldsymbol{\kappa}$, and solution precision parameter ϵ^{dl}.

$\zeta_{\mathrm{low}}^{\mathrm{dl}} \leftarrow 0, \zeta_{\mathrm{up}}^{\mathrm{dl}} \leftarrow 1.$

while $\left|\zeta_{\mathrm{up}}^{\mathrm{dl}} - \zeta_{\mathrm{low}}^{\mathrm{dl}}\right| > \epsilon^{\mathrm{dl}}$ **do**

  $\zeta^{\mathrm{dl}} \leftarrow \frac{\zeta_{\mathrm{up}}^{\mathrm{dl}}+\zeta_{\mathrm{low}}^{\mathrm{dl}}}{2}.$

  Obtain $\{(p_m^{\mathrm{dl}})^*\}_{m\in\mathcal{M}}$ by solving problem (7.37).

  if *Problem* (7.37) *is feasible* **then**

    $\zeta_{\mathrm{up}}^{\mathrm{dl}} \leftarrow \zeta^{\mathrm{dl}}.$

    $(p_m^{\mathrm{dl}})^\star \leftarrow (p_m^{\mathrm{dl}})^*, \forall m \in \mathcal{M}.$

  else

    $\zeta_{\mathrm{low}}^{\mathrm{dl}} \leftarrow \zeta^{\mathrm{dl}}.$

Output: $(\zeta^{\mathrm{dl}})^\star \leftarrow \frac{\zeta_{\mathrm{up}}^{\mathrm{dl}}+\zeta_{\mathrm{low}}^{\mathrm{dl}}}{2}$ and optimal downlink transmit powers $\{(p_m^{\mathrm{dl}})^\star\}_{m\in\mathcal{M}}$.

By denoting $\boldsymbol{q}^{\mathrm{dl}} = \left[\sqrt{p_1^{\mathrm{dl}}}, \sqrt{p_2^{\mathrm{dl}}}, \ldots, \sqrt{p_M^{\mathrm{dl}}}\right]^{\mathsf{T}}$, $\boldsymbol{\Pi}_m = \mathrm{diag}\left(\left[\varpi_{m,0}, \varpi_{m,1}, \ldots, \varpi_{m,M}\right]\right)$, and $\boldsymbol{\vartheta}_m = \left[\vartheta_{m,1}, \vartheta_{m,2}, \ldots, \vartheta_{m,M}\right]^{\mathsf{T}}$ with

$$\vartheta_{m,l} = \begin{cases} \sqrt{\frac{\kappa_m^{\mathrm{dl}}\zeta^{\mathrm{dl}}}{\upsilon_m^2}}, & l = m \\ 0, & l \neq m \end{cases} \tag{7.36}$$

(7.34) can be represented as a set of second-order cone (SOC) constraints. Problem (7.33) is thus equivalent to the following SOCP problem

$$\text{find} \quad \boldsymbol{q}^{\mathrm{dl}} \tag{7.37a}$$

$$* \text{subject to} \quad \left\|[1; \boldsymbol{q}^{\mathrm{dl}}]^{\mathsf{T}}\boldsymbol{\Pi}_m\right\| \leq (\boldsymbol{q}^{\mathrm{dl}})^{\mathsf{T}}\boldsymbol{\vartheta}_m, \forall m \in \mathcal{M} \tag{7.37b}$$

$$*0 \leq q_m \leq \sqrt{P_m^{\mathrm{dl}}}, \forall m \in \mathcal{M} \tag{7.37c}$$

which can be solved by applying convex optimization tools, e.g., CVX [5]. Thus, the optimal downlink transmit power control given ζ^{dl} can be obtained by $(p_m^{\mathrm{dl}})^* = ((q_m^{\mathrm{dl}})^*)^2, \forall m \in \mathcal{M}$, where $(\boldsymbol{q}^{\mathrm{dl}})^* = [(q_1^{\mathrm{dl}})^*, (q_2^{\mathrm{dl}})^*, \ldots, (q_M^{\mathrm{dl}})^*]^{\mathsf{T}}$ is the optimal solution to problem (7.37). Furthermore, by presetting a precision, $\epsilon^{\mathrm{dl}} > 0$, we can leverage the bisection search method to obtain optimal solutions $(\zeta^{\mathrm{dl}})^\star$ and $\{(p_m^{\mathrm{dl}})^\star\}_{m\in\mathcal{M}}$ via checking the feasibility of the solution to problem (7.37), which is summarized in Algorithm 9.

7.4.2 Cooperative Uplink Transmission

For the uplink gradient aggregation, the optimization problem is given by

$$\underset{\zeta_0^{\mathrm{ul}},\{c_m\},\{p_k^{\mathrm{ul}}\}}{\text{minimize}} \quad \zeta_0^{\mathrm{ul}} \tag{7.38a}$$

$$* \text{ subject to} \quad \text{constraints (7.29d), (7.29e), (7.30c), (7.30e).} \tag{7.38b}$$

As the coupling between uplink transmit power levels $\bigcup_{m\in\mathcal{M}}\{p_k^{\mathrm{ul}}\}_{k\in\mathcal{K}_m}$ and receive normalizing factors $\{c_m\}_{m\in\mathcal{M}}$, we can not directly solve problem (7.38). Hence, we first obtain the optimal receive normalizing factors with fixed uplink transmit power levels. To be specific, given $\bigcup_{m\in\mathcal{M}}\{p_k^{\mathrm{ul}}\}_{k\in\mathcal{K}_m}$, problem (7.38) can be decoupled into M subproblems, where the subproblem for cell m is given by

$$\begin{aligned}\underset{c_m>0}{\text{minimize}} \quad & \sum_{k\in\mathcal{K}_m}\left(\frac{|h_k^{\mathrm{ul}}|\sqrt{p_k^{\mathrm{ul}}}}{\sqrt{c_m}} - \upsilon_k\right)^2 + \frac{(\sigma_m^{\mathrm{ul}})^2}{2c_m} \\ & + \sum_{l\in\mathcal{M}\setminus\{m\}}\sum_{k'\in\mathcal{K}_l}\frac{\Re\left\{h_{k',m}^{\mathrm{ul}}(h_{k'}^{\mathrm{ul}})^\dagger\right\}^2 p_{k'}^{\mathrm{ul}}}{|h_{k'}^{\mathrm{ul}}|^2 c_m}.\end{aligned} \tag{7.39}$$

Problem (7.39) is a convex quadratic problem by treating $\frac{1}{\sqrt{c_m}}$ as an optimization variable. The optimal receive normalizing factors can be computed as

$$\begin{aligned} c_m^\star = & \left(\frac{\sum_{k\in\mathcal{K}_m}|h_k^{\mathrm{ul}}|^2 p_k^{\mathrm{ul}} + (\sigma_m^{\mathrm{ul}})^2/2}{\sum_{k\in\mathcal{K}_m}|h_k^{\mathrm{ul}}|\sqrt{p_k^{\mathrm{ul}}}\upsilon_k}\right. \\ & \left.+\frac{\sum_{l\in\mathcal{M}\setminus\{m\}}\sum_{k'\in\mathcal{K}_l}\Re\left\{h_{k',m}^{\mathrm{ul}}(h_{k'}^{\mathrm{ul}})^\dagger\right\}^2 p_{k'}^{\mathrm{ul}}/|h_{k'}^{\mathrm{ul}}|^2}{\sum_{k\in\mathcal{K}_m}|h_k^{\mathrm{ul}}|\sqrt{p_k^{\mathrm{ul}}}\upsilon_k}\right)^2, \\ & \forall m\in\mathcal{M}.\end{aligned} \tag{7.40}$$

On the other hand, substituting (7.40) into problem (7.38), we have

$$\underset{\zeta^{\mathrm{ul}},\{p_k^{\mathrm{ul}}\}}{\text{minimize}} \quad \zeta^{\mathrm{ul}} \tag{7.41a}$$

$$\text{subject to} \quad \left(\sum_{k\in\mathcal{K}_m}\upsilon_k^2 - \kappa_m^{\mathrm{ul}}\zeta^{\mathrm{ul}}\right)\left(\sum_{k\in\mathcal{K}_m}|h_k^{\mathrm{ul}}|^2 p_k^{\mathrm{ul}} + \frac{(\sigma_m^{\mathrm{ul}})^2}{2}\right.$$

$$+ \sum_{l \in \mathcal{M} \setminus \{m\}} \sum_{k' \in \mathcal{K}_l} \frac{\Re\left\{h_{k',m}^{\mathrm{ul}} (h_{k'}^{\mathrm{ul}})^{\dagger}\right\}^2 p_{k'}^{\mathrm{ul}}}{|h_{k'}^{\mathrm{ul}}|^2} \Bigg)$$

$$\leq \left(\sum_{k \in \mathcal{K}_m} |h_k^{\mathrm{ul}}| \sqrt{p_k^{\mathrm{ul}}} \upsilon_k \right)^2, \ \forall m \in \mathcal{M} \quad (7.41b)$$

$$0 \leq p_k^{\mathrm{ul}} \leq P_k^{\mathrm{ul}}, \ \forall k \in \mathcal{K}_m, \ \forall m \in \mathcal{M} \quad (7.41c)$$

$$\zeta^{\mathrm{ul}} \geq 0 \quad (7.41d)$$

where $\zeta^{\mathrm{ul}} = \frac{\zeta_0^{\mathrm{ul}}}{L}$ and $\kappa_m^{\mathrm{ul}} = \frac{\kappa_m K_m^2}{\eta_m}$. We can solve this problem in a way similar to solving problem (7.32). Specifically, given ζ^{ul}, problem (7.41) becomes

$$\text{find} \quad \{p_k^{\mathrm{ul}}\} \quad (7.42a)$$

$$* \text{subject to} \quad \Xi_m \Bigg(\sum_{k \in \mathcal{K}_m} |h_k^{\mathrm{ul}}|^2 p_k^{\mathrm{ul}} + \frac{(\sigma_m^{\mathrm{ul}})^2}{2}$$

$$+ \sum_{l \in \mathcal{M} \setminus \{m\}} \sum_{k' \in \mathcal{K}_l} \frac{\Re\left\{h_{k',m}^{\mathrm{ul}} (h_{k'}^{\mathrm{ul}})^{\dagger}\right\}^2 p_{k'}^{\mathrm{ul}}}{|h_{k'}^{\mathrm{ul}}|^2} \Bigg)$$

$$\leq \left(\sum_{k \in \mathcal{K}_m} |h_k^{\mathrm{ul}}| \sqrt{p_k^{\mathrm{ul}}} \upsilon_k \right)^2, \ \forall m \in \mathcal{M} \quad (7.42b)$$

$$*0 \leq p_k^{\mathrm{ul}} \leq P_k^{\mathrm{ul}}, \ \forall k \in \mathcal{K}_m, \ \forall m \in \mathcal{M} \quad (7.42c)$$

where $\Xi_m = \sum_{k \in \mathcal{K}_m} \upsilon_k^2 - \kappa_m^{\mathrm{ul}} \zeta^{\mathrm{ul}} \geq 0$. By defining $\boldsymbol{\gamma}_m = [\gamma_{m,1}, \gamma_{m,2}, \ldots, \gamma_{m,K_{\mathrm{tot}}}]$, $\forall m \in \mathcal{M}$, with $K_{\mathrm{tot}} = \sum_{m \in \mathcal{M}} K_m$ and

$$\gamma_{m,k} = \begin{cases} |h_k^{\mathrm{ul}}|, & k \in \mathcal{K}_m \\ \frac{\Re\left\{h_{k,m}^{\mathrm{ul}} (h_k^{\mathrm{ul}})^{\dagger}\right\}}{|h_k^{\mathrm{ul}}|}, & k \in \mathcal{K}_l, \ \forall l \in \mathcal{M} \setminus \{m\} \end{cases} \quad (7.43)$$

constraints (7.42b) can be reformulated as

$$\sqrt{\Xi_m \left(\sum_{l \in \mathcal{M}} \sum_{k \in \mathcal{K}_l} \gamma_{m,k}^2 p_k^{\mathrm{ul}} + \frac{(\sigma_m^{\mathrm{ul}})^2}{2} \right)} \leq \sum_{k \in \mathcal{K}_m} |h_k^{\mathrm{ul}}| \sqrt{p_k^{\mathrm{ul}}} \upsilon_k,$$

$$\forall m \in \mathcal{M}. \quad (7.44)$$

Algorithm 10: Cooperative uplink optimization for problem (7.38)

Input: Standard deviations of local gradients $\{\upsilon_k\}$, profiling vector $\boldsymbol{\kappa}$, and solution precision parameter ϵ^{ul}.

$\zeta_{\text{low}}^{\text{ul}} \leftarrow 0$, $\zeta_{\text{up}}^{\text{ul}} \leftarrow \min_{m\in\mathcal{M}} \frac{\sum_{k\in\mathcal{K}_m}\upsilon_k^2}{\kappa_m^{\text{ul}}}$.

while $\left|\zeta_{\text{up}}^{\text{ul}} - \zeta_{\text{low}}^{\text{ul}}\right| > \epsilon^{\text{ul}}$ **do**

  $\zeta^{\text{ul}} \leftarrow \frac{\zeta_{\text{up}}^{\text{ul}}+\zeta_{\text{low}}^{\text{ul}}}{2}$.

  Obtain $\bigcup_{m\in\mathcal{M}}\{(p_m^{\text{ul}})^*\}_{k\in\mathcal{K}_m}$ by solving problem (7.37).

  if *Problem* (7.37) *is feasible* **then**

    $\zeta_{\text{up}}^{\text{ul}} \leftarrow \zeta^{\text{ul}}$.

    $(p_k^{\text{ul}})^\star \leftarrow (p_k^{\text{ul}})^*$, $\forall k \in \mathcal{K}_m$, $\forall m \in \mathcal{M}$.

  else

    $\zeta_{\text{low}}^{\text{ul}} \leftarrow \zeta^{\text{ul}}$.

Obtain $\{c_m^\star\}$ based on (7.40).

Output: $(\zeta^{\text{ul}})^\star \leftarrow \frac{\zeta_{\text{up}}^{\text{ul}}+\zeta_{\text{low}}^{\text{ul}}}{2}$, optimal uplink transmit powers $\bigcup_{m\in\mathcal{M}}\{(p_k^{\text{ul}})^\star\}_{k\in\mathcal{K}_m}$, and optimal receive normalizing factors $\{c_m^\star\}_{m\in\mathcal{M}}$.

Then, problem (7.42) can be rewritten as the following SOCP problem:

$$\text{find} \quad \boldsymbol{q}^{\text{ul}} \tag{7.45a}$$

$$* \text{ subject to} \quad \sqrt{\Xi_m}\left\|[1; \boldsymbol{q}^{\text{ul}}]^{\mathsf{T}}\boldsymbol{\Gamma}_m\right\| \le (\boldsymbol{q}^{\text{ul}})^{\mathsf{T}}\boldsymbol{\psi}_m, \ \forall m \in \mathcal{M} \tag{7.45b}$$

$$*0 \le q_k^{\text{ul}} \le \sqrt{P_k^{\text{ul}}}, \ \forall k \in \mathcal{K}_m, \ \forall m \in \mathcal{M} \tag{7.45c}$$

where $\boldsymbol{q}^{\text{ul}} = \left[\sqrt{p_1^{\text{ul}}}, \sqrt{p_2^{\text{ul}}}, \ldots, \sqrt{p_{K_{\text{tot}}}^{\text{ul}}}\right]^{\mathsf{T}}$, $\boldsymbol{\Gamma}_m = \text{diag}\left(\left[\sigma_m^{\text{ul}}/\sqrt{2}, \boldsymbol{\gamma}_m\right]\right)$, and $\boldsymbol{\psi}_m = \left[\psi_{m,1}, \psi_{m,2}, \ldots, \psi_{m,K_{\text{tot}}}\right]^{\mathsf{T}}$ with

$$\psi_{m,k} = \begin{cases} |h_k^{\text{ul}}|\upsilon_k, & k \in \mathcal{K}_m \\ 0, & k \in \mathcal{K}_l, \ \forall l \in \mathcal{M}\setminus\{m\}. \end{cases} \tag{7.46}$$

Hence, the optimal uplink transmit power $(p_k^{\text{ul}})^* = ((q_k^{\text{ul}})^*)^2$ for all $k \in \mathcal{K}_m$ and $m \in \mathcal{M}$ can be obtained using convex optimization tools by solving problem (7.45), given ζ^{ul}. Furthermore, the bisection search method can be employed to find the optimal solutions $(\zeta^{\text{ul}})^\star$ and $\bigcup_{m\in\mathcal{M}}\{(p_k^{\text{ul}})^\star\}_{k\in\mathcal{K}_m}$ for problem (7.41) with a given solution precision ϵ^{ul}. The optimal receive normalizing factors $\{c_m^\star\}_{m\in\mathcal{M}}$ can be obtained by substituting $\bigcup_{m\in\mathcal{M}}\{(p_k^{\text{ul}})^\star\}_{k\in\mathcal{K}_m}$ into Eq. (7.40). Algorithm 10 summarizes the uplink transmission optimization process for problem (7.38).

7.5 Simulation Results

7.5.1 Simulation Setup

Consider a four-cell wireless network, where the BSs locate at $(0, 0)$, $(40, 0)$, $\left(20, 20\sqrt{3}\right)$, and $\left(20, -20\sqrt{3}\right)$ meters, respectively. The devices in each cell are uniformly and randomly distributed in a circle of radius $\varrho \in [1, 20]$ meters centered at their home BS. The number of devices located in each cell is set to $K_m = \bar{K} = 10$, $\forall m \in \mathcal{M}$. All channel coefficients are modeled as [12]

$$h = \rho^{-\alpha/2}\left(\sqrt{\frac{\beta}{1+\beta}}h_{\text{LoS}} + \sqrt{\frac{1}{1+\beta}}h_{\text{NLoS}}\right), \tag{7.47}$$

which are independent over different rounds, where ρ denotes the distance between the transmitter and the receiver, $\alpha = 2.5$ denotes the pathloss exponent, $\beta = 5$ (dB) represents the Rician factor, h_{LoS} denotes the line-of-sight (LoS) component, and $h_{\text{NLoS}} \sim \mathcal{CN}(0, 1)$ denotes the non-line-of-sight (NLoS) exponent. Set the maximum downlink transmit power budgets of four BSs as $P_1^{\text{dl}} = 40$ dBm, $P_2^{\text{dl}} = 30$ dBm, $P_3^{\text{dl}} = 30$ dBm, and $P_4^{\text{dl}} = 40$ dBm, respectively. The maximum transmit power budgets of devices in each cell are set to $P_k^{\text{ul}} = 15$ dBm, $\forall k \in \{\sum_{l=1}^{m-1} K_l + 1, \ldots, \sum_{l=1}^{m-1} K_l + \lfloor\frac{\bar{K}}{2}\rfloor\}$, and $P_k^{\text{ul}} = 30$ dBm, $\forall k \in \{\sum_{l=1}^{m-1} K_l + \lfloor\frac{\bar{K}}{2}\rfloor + 1, \ldots, \sum_{l=1}^{m-1} K_l + \bar{K}\}$, $\forall m \in \mathcal{M}$, where $\lfloor x \rfloor$ is the floor function that returns the greatest integer less than or equal to $x \in \mathbb{R}$. Besides, set the noise power as $(\sigma_m^{\text{ul}})^2 = (\sigma_k^{\text{dl}})^2 = -110$ dBm, $\forall k \in \mathcal{K}_m$, $\forall m \in \mathcal{M}$, and the solution precision is set to $\epsilon^{\text{dl}} = \epsilon^{\text{ul}} = 10^{-9}$.

7.5.1.1 Learning Model Setting

The multinomial logistic regression technique is utilized for training the learning models. In the multi-cell FEEL system under consideration, we employ the MNIST [7] and Fashion-MNIST [15] datasets. Each cell in the system is responsible for different learning tasks based on its assigned dataset. To be specific, cell 1 is assigned the labeled with $0 \sim 4$ from the MNIST dataset, while cell 2 is assigned the data labeled with $5 \sim 9$ from the same dataset. On the other hand, cell 3 handles the data labeled with $0 \sim 4$ from the Fashion-MNIST dataset, and 4 takes care of the data labeled with $5 \sim 9$ the Fashion-MNIST dataset. Within each cell, we follow a three-step process: first, we sort the dataset based on the labels it contains; then, we divide it into $\bar{K}$ shards; finally, we assign one shard to each device without replacement. The learning rates for the four cells are set as follows: $\eta_1 = \eta_2 = 0.1$ and $\eta_3 = \eta_4 = 0.01$.

7.5.1.2 Baseline Schemes

We consider the following transmission schemes for comparison.

- **Benchmark/DL-Free / UL-Free**: The scenario where both downlink model dissemination and uplink gradient aggregation are conducted without errors serves as the *Benchmark* scheme. Error-free downlink and uplink transmissions are referred to as *DL-Free* and *UL-Free*, respectively.
- **DL-Opt/UL-Opt**: *DL-Opt* refers to the strategy that allocates resources for downlink transmission using Algorithm 9, while *UL-Opt* refers to the strategy that allocates resources for uplink transmission using Algorithm 10.
- **UL-IgnInter**: Each cell independently optimizes its uplink transmission by ignoring the inter-cell interference.
- **UL-MaxInter**: Each cell optimizes its uplink transmission by assuming maximum inter-cell interference from other cells.
- **DL-Full/UL-Full**: In the *DL-Full* scheme, full-power transmission is applied by setting $p_m^{\mathrm{dl}} = P_m^{\mathrm{dl}}$ for all $m \in \mathcal{M}$. In the *UL-Full* scheme, it is applied by setting $p_k^{\mathrm{ul}} = P_k^{\mathrm{ul}}$ for all $k \in \mathcal{K}_m$ and $m \in \mathcal{M}$.

7.5.2 Performance Evaluation

7.5.2.1 Two-Cell Network

Initially, we assess the effectiveness of our novel collaborative multi-cell FEEL optimization framework in a dual-cell network. Specifically, only cell 1 and cell 2 are operational, and we utilize a profiling vector $\boldsymbol{\kappa} = \left[\frac{1}{2}, \frac{1}{2}\right]$ for evaluation purposes.

Figure 7.3 presents the learning performance across different rounds for various downlink transmission schemes, assuming error-free uplink gradient aggregation. It is worth noting the test accuracy achieved by *DL-Opt* closely resemble those obtained by *Benchmark* in both cells. This finding highlights the effectiveness of our proposed cooperative downlink transmission approach in mitigating downlink dissemination errors between the two cells while maintaining satisfactory learning performance. When examining cell 1 and cell 2, we observe that *DL-Full* exhibits similar training loss and test accuracy to *Benchmark* in cell 1, but displays larger performance disparities compared to *Benchmark* in cell 2. This discrepancy suggests that the absence of interference management results in significant degradation of downlink transmission in cell 2 due to inter-cell interference originating from cell 1. The magnitude of downlink dissemination errors at each device follows a pattern where it increases with the transmission power of non-associated base stations (BSs) and decreases with the transmission power of the home BS, as indicated by Eq. (7.7). Since the maximumlink transmit power budget of cell 1 surpasses that of cell 2, employing downlink full-power transmission alone leads to inadequate suppression

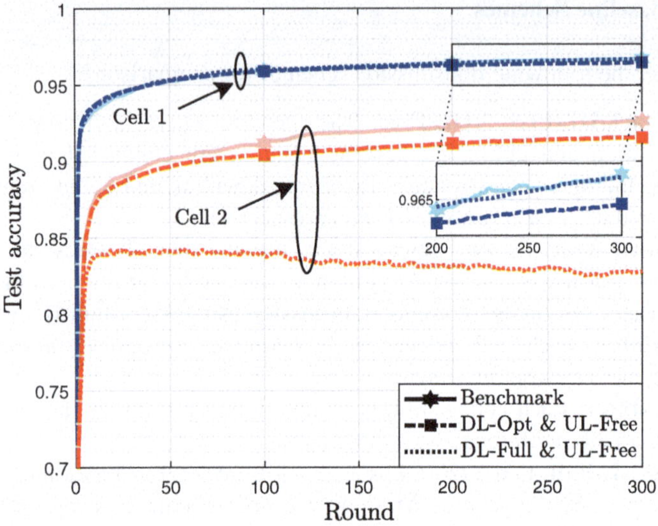

Fig. 7.3 Test accuracy versus the number of rounds

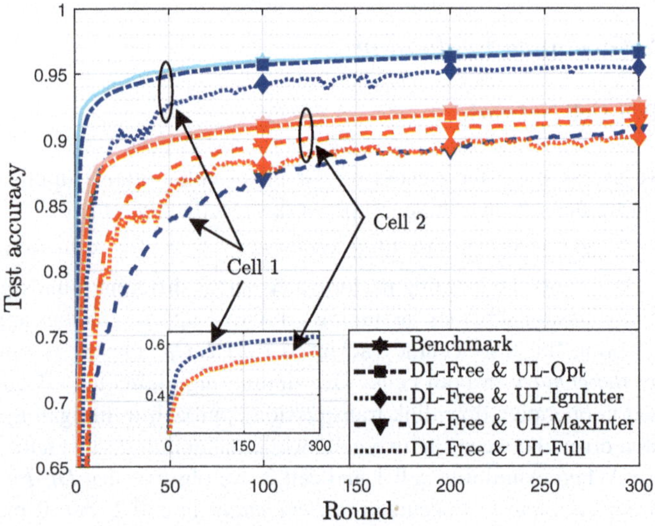

Fig. 7.4 Test accuracy versus the number of rounds

of inter interference for devices in cell 2, consequently diminishing the accuracy of the received global model at each device.

Figure 7.4 presents the learning performance across different rounds for various uplink transmission schemes, assuming error-free downlink model dissemination. It is worth noting that our proposed cooperative uplink transmission, referred to as *UL-Opt*, achieves comparable results to the *Benchmark* in terms of test accuracy.

This outcome demonstrates the effectiveness of our approach in balancing uplink aggregation errors between two cells while maintaining near-optimal learning performance. Unlike the downlink model dissemination, the transmission design in uplink gradient aggregation requires achieving magnitude alignment at the receiver while suppressing inter-cell interference to accomplish the desired function computation [17]. However, *UL-IgnInter* and *UL-MaxInter* exhibit less accurate power control compared to *UL-Opt*. Consequently, leads to larger distortion of the aggregated local gradients at the base station (BS) and subsequently results in inferior learning performance in both cells. Specifically, *UL-IgnInter* suffers from severe inter-cell interference during the training process due to its negligence of inter-cell interference. As channel conditions change independently over different communication rounds, the inter-cell interference may vary significantly, leading to substantial fluctuations in the training loss of *UL-IgnInter*. On the other hand, *UL-MaxInter* assumes the worst effect of inter interference, resulting in a relatively smooth curve compared to *UL-IgnInter*. Moreover, the performance of *UL-Full* is heavily dependent on factors such as receiver noise, channel fading, and inter-cell interference since it lacks power control. Consequently, it exhibits much worse learning performance than other baseline schemes. Furthermore, due to the randomness of receiver noise and channel fading, the convergence of federated edge learning (FEEL) training using *UL-Full* is not guaranteed. As a result, some experiments yield an increasing training loss and a non-increasing test accuracy, while others show a slight decrease in training loss and an increase in test accuracy.

7.5.2.2 Four-Cell Network

We next assess the average learning performance over multiple cells of our proposed cooperative multi-cell FEEL optimization framework in multi-cell wireless networks. Specifically, we consider the active cells as $1 \sim M$, and we configure the profiling vector to $\boldsymbol{\kappa} = \left[\frac{1}{M}, \frac{1}{M}, \ldots, \frac{1}{M}\right]$.

Figure 7.5 present the average learning performance as a function of the number of rounds for different transmission schemes in a four-cell network. It is worth noting the proposed cooperative transmission design, namely *DL-Opt & UL-Opt*, surpasses other baseline schemes that lack interference management in either downlink or uplink transmission in terms of average training loss and test accuracy. This finding highlights the effectiveness of our approach in achieving balanced learning performance in multi wireless networks. Interestingly, an increasing number of rounds, the learning performance of *DL-Opt & UL-MaxInter* gradually converges to and even surpasses that of *DL-Opt & UL-IgnInter*. Consequently, inter-cell interference emerges as a significant limiting factor for performance improvement, underscoring the importance of interference management in multi-cell wireless networks. Furthermore, similar to the behavior observed in Fig. 7.5 for *DL-Free & UL-IgnInter*, the performance of *DL-Opt & UL-IgnInter* is primarily constrained by

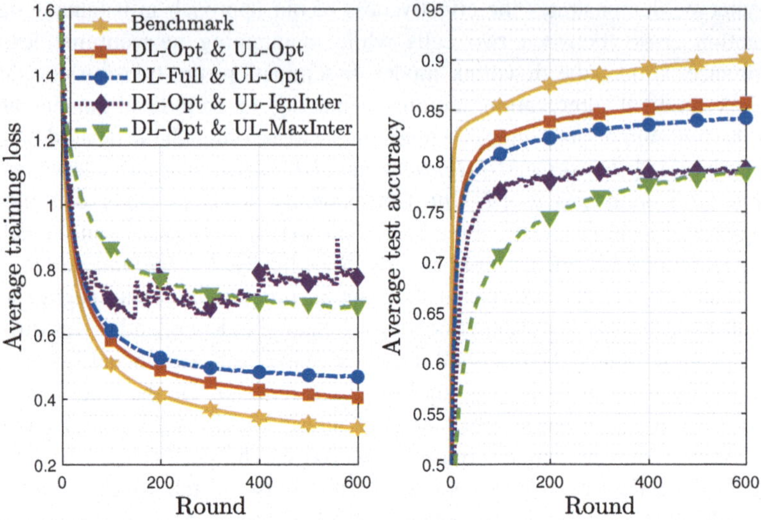

Fig. 7.5 Average learning performance in a four-cell network

uplink transmission, resulting in relatively large fluctuations in training loss when inter-cell interference is ignored.

7.6 Summary

In this chapter, we presented over-the-air FEEL in a multi-cell wireless network with multiple FEEL tasks. To evaluate the effectiveness of the learning process, we analyzed the convergence of over-the-air FEEL, which shows the negative impact of distorted downlink and uplink communications, resulting in a significant gap that obstructs the convergence of FEEL. We employed a profiling technique to characterize the Pareto boundary of the gap region and formulated an optimization problem to minimize the cumulative gaps induced by errors across all cells. By leveraging the coordinated efforts of multiple cells, our proposed cooperative algorithm consistently outperformed non-cooperative baseline schemes, leading to significantly improved average learning performance.

References

1. M. M. Amiri and D. Gündüz. Federated learning over wireless fading channels. *IEEE Trans. Wireless Commun.*, 19(5):3546–3557, May 2020.
2. M. M. Amiri and D. Gündüz. Machine learning at the wireless edge: Distributed stochastic gradient descent over-the-air. *IEEE Trans. Signal Process.*, 68:2155–2169, Mar. 2020.

3. Q. An, Y. Zhou, Z. Wang, H. Shan, Y. Shi, and M. Bennis. Online optimization for over-the-air federated learning with energy harvesting. *IEEE Trans. Wireless Commun.*, 2023.
4. X. Cao, G. Zhu, J. Xu, and K. Huang. Cooperative interference management for over-the-air computation networks. *IEEE Trans. Wireless Commun.*, 20(4):2634–2651, Apr. 2021.
5. M. Grant and S. Boyd. CVX: Matlab software for disciplined convex programming, version 2.1. http://cvxr.com/cvx, Mar. 2014.
6. E. A. Jorswieck, E. G. Larsson, and D. Danev. Complete characterization of the Pareto boundary for the MISO interference channel. *IEEE Trans. Signal Process.*, 56(10):5292–5296, Oct. 2008.
7. Y. Lecun, L. Bottou, Y. Bengio, and P. Haffner. Gradient-based learning applied to document recognition. *Proc. IEEE*, 86(11):2278–2324, Nov. 1998.
8. Z. Lin, X. Li, V. K. N. Lau, Y. Gong, and K. Huang. Deploying federated learning in large-scale cellular networks: Spatial convergence analysis. *IEEE Trans. Wireless Commun.*, 21(3):1542–1556, Mar. 2022.
9. Y. Pan, Z. Wang, L. Wu, and Y. Zhou. Irs-assisted digital over-the-air federated learning. In *GLOBECOM 2023–2023 IEEE Global Communications Conference*, pages 3276–3281. IEEE, 2023.
10. M. Salehi and E. Hossain. Federated learning in unreliable and resource-constrained cellular wireless networks. *IEEE Trans. Commun.*, 69(8):5136–5151, Aug. 2021.
11. Y. Shi, S. Xia, Y. Zhou, Y. Mao, C. Jiang, and M. Tao. Vertical federated learning over cloud-ran: Convergence analysis and system optimization. *IEEE Transactions on Wireless Communications*, 23(2):1327–1342, 2023.
12. D. Tse and P. Viswanath. *Fundamentals of wireless communication*. Cambridge Univ. Press, Cambridge, U.K., Jul. 2005.
13. Z. Wang, Y. Zhou, Y. Shi, and W. Zhuang. Interference management for over-the-air federated learning in multi-cell wireless networks. *IEEE Journal on Selected Areas in Communications*, 40(8):2361–2377, 2022.
14. Z. Wang, Y. Zhou, Y. Zou, Q. An, Y. Shi, and M. Bennis. A graph neural network learning approach to optimize ris-assisted federated learning. *IEEE Transactions on Wireless Communications*, 22(9):6092–6106, 2023.
15. H. Xiao, K. Rasul, and R. Vollgraf. Fashion-MNIST: a novel image dataset for benchmarking machine learning algorithms. 2017.
16. R. Zhang and S. Cui. Cooperative interference management with MISO beamforming. *IEEE Trans. Signal Process.*, 58(10):5450–5458, Oct. 2010.
17. G. Zhu, J. Xu, K. Huang, and S. Cui. Over-the-air computing for wireless data aggregation in massive IoT. *IEEE Wireless Commun.*, 28(4):57–65, Aug. 2021.
18. H. Zhu, Y. Zhou, H. Qian, Y. Shi, X. Chen, and Y. Yang. Online client selection for asynchronous federated learning with fairness consideration. *IEEE Transactions on Wireless Communications*, 22(4):2493–2506, 2022.
19. Y. Zou, Z. Wang, X. Chen, H. Zhou, and Y. Zhou. Knowledge-guided learning for transceiver design in over-the-air federated learning. *IEEE Transactions on Wireless Communications*, 22(1):270–285, 2022.

Part IV
Security and Privacy

In this part, we delve into the security and privacy challenges of FEEL, proposing differentially private model aggregation and Byzantine-resilient model aggregation schemes.

Chapter 8
Differentially Private Federated Edge Learning

8.1 Overview

Recent studies on FEEL [4, 8, 20] have demonstrated that transmitting model updates, such as local gradients, can still result in privacy concerns. To address this, researchers have introduced a rigorous mathematical framework called differential privacy (DP) [3, 5] to quantify privacy disclosure. Several privacy-preserving techniques have been developed, including the injection of random perturbations following specific distributions like Gaussian [3], Laplacian [16], and Binomial [1]. Notably, an alternative FEEL scheme based on AirComp was proposed in [10], which leverages wireless channel noise to safeguard user privacy at no additional cost. The authors of [7] further established the inherent anonymity of AirComp, concealing individual local updates within the collective to ensure robust privacy while minimizing the need for artificial noise. Another approach presented in [13] introduces perturbations to a subset of edge devices, benefiting the entire system. However, it is important to note that ensuring privacy in FEEL often comes at the expense of learning accuracy due to the reduction in signal-to-noise ratio (SNR) [10, 18], necessitating a trade-off between accuracy and privacy.

In order to address the delicate balance between privacy and accuracy, this chapter leverages both RIS and AirComp to empower FEEL with DP guarantenn. We propose a practical transmission scheme for local model updates, wherein the high-dimensional model updates (such as deep neural network models) are distributed across multiple communication blocks within a single learning round [9]. Through extensive convergence analysis and system optimization, our findings demonstrate that the RIS-enabled FEEL system not only enhances the system's SNR and amplifies the received signal power, but also establishes remarkable learning accuracy while effectively meeting the privacy requirements.

Y. Zhou et al., *Federated Edge Learning*, Wireless Networks,
https://doi.org/10.1007/978-3-031-96649-1_8

8.2 System Model

8.2.1 Distributed Federated Edge Learning

Consider an RIS-enabled FEEL system consisting of one single-antenna edge server, K single-antenna edge devices indexed by set $\mathcal{K} = \{1, \ldots, K\}$, and an RIS equipped with N passive reflecting elements, as depicted in Fig. 8.1. Mathematically, the problem of FEEL can be represented as

$$\min_{\boldsymbol{\theta}} F(\boldsymbol{\theta}) = \frac{1}{\sum_{k=1}^{K} D_k} \sum_{k=1}^{K} D_k F_k(\boldsymbol{\theta}), \tag{8.1}$$

where $\mathcal{D}_k$ denotes the local dataset of device k with $D_k = |\mathcal{D}_k|$ data samples, $F_k(\boldsymbol{\theta})$ denotes the loss function of device k defined as

$$F_k(\boldsymbol{\theta}) = \frac{1}{D_k} \sum_{(\boldsymbol{x}, y) \in \mathcal{D}_k} f(\boldsymbol{x}, y; \boldsymbol{\theta}), \tag{8.2}$$

where $f(\boldsymbol{x}, y; \boldsymbol{\theta})$ represents the sample-wise loss function. In this chapter, we assume that all local data are i.i.d. and the local datasets have the same size [19], i.e., $D_k = D, \ \forall k \in \mathcal{K}$.

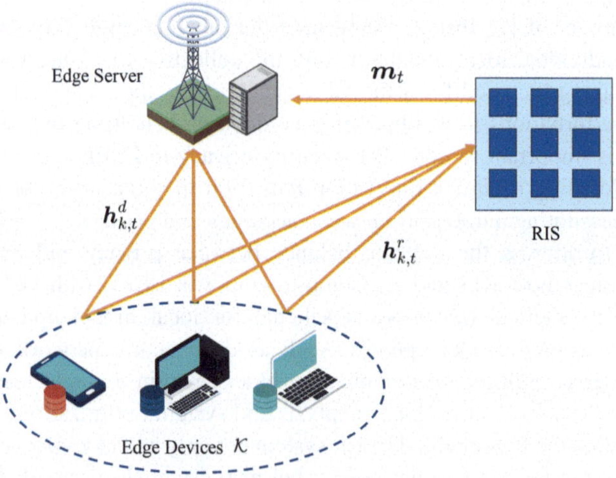

Fig. 8.1 An RIS-enabled wireless FEEL system

Full-batch gradient descent is adopted to solve the above problem. Each learning round consists of the following procedures:

- *Model Dissemination*: The edge server disseminates the latest global model $\boldsymbol{\theta}_t$ to all edge devices.
- *Local update*: Each edge device k computes its local update $\boldsymbol{g}_{k,t}$ with respect to its local dataset $\mathcal{D}_k$ based on $\boldsymbol{\theta}_t$, i.e.,

$$\boldsymbol{g}_{k,t} = \nabla F_k\left(\boldsymbol{\theta}_t\right) \in \mathbb{R}^d. \tag{8.3}$$

- *Model Aggregation*: All the edge devices upload their gradients to the edge server. Then the edge server updates $\boldsymbol{\theta}_{t+1}$, i.e.,

$$\boldsymbol{\theta}_{t+1} = \boldsymbol{\theta}_t - \lambda \hat{\boldsymbol{g}}_t, \tag{8.4}$$

where λ represents the learning rate and $\hat{\boldsymbol{g}}_t$ denotes the global gradient $\boldsymbol{g}_t = \nabla F\left(\boldsymbol{\theta}_t\right) = \frac{1}{K}\sum_{k=1}^{K} \boldsymbol{g}_{k,t}$.

8.2.2 Channel Model for RIS-Enabled FEEL Systems

The accuracy of model aggregation in the FEEL system can be significantly influenced by unfavorable channel propagations and power limitations of individual edge devices, leading to a decrease in learning accuracy. To tackle this challenge, a solution is implemented by deploying an RIS with N reflecting elements. This RIS aims to improve the channel conditions of the links connecting the edge server and all edge devices [11].

For the simplicity of the theoretical analysis, we assume the downlink channels to be noise-free [2, 14] and only focus on the uplink fading channels for aggregating the local models [15]. Given that the model parameters dimension d exceeds the channel coherence length e [9], each edge device uploads its local update sequentially over several consecutive communication blocks. Specifically, in the t-th learning round, the whole update message $\boldsymbol{x}_{k,t} = [\boldsymbol{x}_{k,t}^T(1), \ldots, \boldsymbol{x}_{k,t}^T(I)]^T \in \mathbb{C}^d$ that is a function of $\boldsymbol{g}_{k,t}$ is evenly divided into $I = \lceil d/e \rceil$ e-dimensional vectors denoted by $\{\boldsymbol{x}_{k,t}(i) \in \mathbb{C}^e\}$. Then $\{\boldsymbol{x}_{k,t}(i)\}$ are sequentially transmitted across I communication blocks. We assume $I = d/e$ for simplicity.

For the i-th communication block in learning round t, let $h_{k,t}^d(i) \in \mathbb{C}$, $\boldsymbol{h}_{k,t}^r(i) \in \mathbb{C}^N$, and $\boldsymbol{m}_t(i) \in \mathbb{C}^N$ denote the channel responses from edge device k to the edge server, from edge device k to RIS, and from RIS to the edge server, respectively. Channel coefficients are assumed to be independent across communication blocks and constant within a single communication block. Additionally, the RIS phase-shift matrix is denoted as $\boldsymbol{\Theta}_t(i) = \mathrm{diag}(e^{j\phi_{1,t}(i)}, \ldots, e^{j\phi_{N,t}(i)}) \in \mathbb{C}^{N\times N}$, where $\phi_{n,t}(i) \in [0, 2\pi)$, $n \in \{1, \ldots, N\}$. The composite channel response of the k-th edge device is thus denoted as $h_{k,t}(i) = \boldsymbol{m}_t^T(i)\boldsymbol{\Theta}_t(i)\boldsymbol{h}_{k,t}^r(i) + h_{k,t}^d(i)$.

During the learning round t, the aggregated signal at the edge server in the t-th communication block is given by

$$\boldsymbol{r}_t(i) = \sum_{k=1}^{K} h_{k,t}(i)\,\boldsymbol{x}_{k,t}(i) + \boldsymbol{w}_t(i), \tag{8.5}$$

where $\boldsymbol{r}_t = [\boldsymbol{r}_t^T(1), \ldots, \boldsymbol{r}_t^T(I)]^T$ is the aggregated signal for the t-th learning round and $\boldsymbol{w}_t(i) \sim \mathcal{CN}(\boldsymbol{0}, N_0\boldsymbol{I}_e)$ is the additive white Gaussian noise (AWGN).

8.2.3 *Differential Privacy*

Although the edge devices never share sensitive raw data. According to [4], the transmission of model updates $\boldsymbol{g}_{k,t}$ can potentially disclose statistical information about the local dataset, necessitating additional measures to enhance privacy protection. We assume that the edge server, while trustworthy, possesses a curiosity to infer the local information of each edge device based on the received signals $\boldsymbol{r} = \{\boldsymbol{r}_t\}_{t=1}^{T}$. To address this concern, we introduce the concept of neighboring datasets and employ Differential Privacy (DP). A dataset $\mathcal{D}_k = \{\boldsymbol{x}_1, \ldots, \boldsymbol{x}_n\} \in \mathcal{X}^n$ consists of n data points from $\mathcal{X}$. Two datasets, $\mathcal{D}_k = \{\boldsymbol{x}_1, \ldots, \boldsymbol{x}_n\}$ and $\mathcal{D}_k' = \{\boldsymbol{x}_1', \ldots, \boldsymbol{x}_n'\}$, are considered neighboring if they differ by only one element. In other words, there exists an index $i \in [n]$ such that $\boldsymbol{x}_i \neq \boldsymbol{x}_i'$, while $\boldsymbol{x}_j = \boldsymbol{x}_j'$ for all $i \neq j$. Consequently, we present the subsequent definition of DP, as described in [3, 5].

Definition 8.1 For $\epsilon > 0$, $0 \leq \delta \leq 1$, and an arbitrary protocol $\mathcal{M} : \mathcal{X}^n \to \mathcal{Y}$, protocol $\mathcal{M}$ is (ϵ, δ)-differentially private (in short, (ϵ, δ)-DP), if for any two possible neighboring datasets $\mathcal{D}_k, \mathcal{D}_k' \in \mathcal{X}^n$ and any subset $\Lambda \subseteq \mathcal{Y}$, the following inequality holds:

$$\Pr[\mathcal{M}(\mathcal{D}_k) \in \Lambda] \leq \mathrm{e}^{\epsilon} \Pr[\mathcal{M}(\mathcal{D}_k') \in \Lambda] + \delta. \tag{8.6}$$

Note that the case of $\delta = 0$ is called pure ϵ-DP.

To achieve the privacy level ϵ, we uniformly perturb the signal with artificial noise $\boldsymbol{n}_{k,t} = [\boldsymbol{n}_{k,t}^T(1), \ldots, \boldsymbol{n}_{k,t}^T(I)]^T \in \mathbb{C}^d$, where $\boldsymbol{n}_{k,t}(i) \sim \mathcal{CN}(\boldsymbol{0}, \sigma_{k,t}^2(i)\,\boldsymbol{I}_e)$, to each communication block in the t-th learning round, producing a noisy pre-processed local update at the edge device k as

$$\boldsymbol{s}_{k,t}(i) = D_k \boldsymbol{g}_{k,t}(i) + \boldsymbol{n}_{k,t}(i). \tag{8.7}$$

8.2.4 Model Aggregation via Over-the-Air Computation

With AirComp, all edge devices concurrently send their gradients $\{\boldsymbol{g}_{k,t}(i)\}_{k=1}^{K}$ over the same time-frequency resources block.

To compensate the non-uniform fading of channel, we adopt the following transmitter scheme:

$$\boldsymbol{x}_{k,t}(i) = \alpha_{k,t}(i)\, \boldsymbol{s}_{k,t}(i) = \sqrt{\eta_t(i)}\, \frac{(h_{k,t}(i))^H}{|h_{k,t}(i)|^2}\, \boldsymbol{s}_{k,t}(i), \tag{8.8}$$

where $\alpha_{k,t}(i) \in \mathbb{C}$ denotes the transmit scalar given by $\alpha_{k,t}(i) = \sqrt{\eta_t(i)}/h_{k,t}(i)$ and $\eta_t(i)$ denotes the power scaling factor. Given the limited transmit energy,we have

$$\mathbb{E}\left(||\boldsymbol{x}_{k,t}(i)||^2\right) = \mathbb{E}\left(||\alpha_{k,t}(i)\boldsymbol{s}_{k,t}(i)||^2\right) \le P_0,\ \forall k, i, t. \tag{8.9}$$

Therefore, based on (8.7) and (8.8), the aggregated signal $\boldsymbol{r}_t(i)$ defined in (8.5) can be rewritten as

$$\boldsymbol{r}_t(i) = \sqrt{\eta_t(i)} \sum_{k=1}^{K} D_k \boldsymbol{g}_{k,t}(i) + \sqrt{\eta_t(i)} \sum_{k=1}^{K} \boldsymbol{n}_{k,t}(i) + \boldsymbol{w}_t(i). \tag{8.10}$$

Then, the edge server computes the estimated global update $\hat{\boldsymbol{g}}_t(i)$ by

$$\begin{aligned} \hat{\boldsymbol{g}}_t(i) &= \frac{1}{KD\sqrt{\eta_t(i)}} \mathrm{Re}\{\boldsymbol{r}_t(i)\} \\ &= \boldsymbol{g}_t(i) + \frac{1}{KD} \sum_{k=1}^{K} \mathrm{Re}\{\boldsymbol{n}_{k,t}(i)\} + \frac{\mathrm{Re}\{\boldsymbol{w}_t(i)\}}{KD\sqrt{\eta_t(i)}}, \end{aligned} \tag{8.11}$$

where $\boldsymbol{g}_t(i) = \sum_{k=1}^{K} \boldsymbol{g}_{k,t}(i)/K$. After I rounds of model aggregation, the resulting $\hat{\boldsymbol{g}}_t = [\hat{\boldsymbol{g}}_t^T(1), \ldots, \hat{\boldsymbol{g}}_t^T(I)]^T$ is an unbiased estimation of the global gradient $\boldsymbol{g}_t = [\boldsymbol{g}_t^T(1), \ldots, \boldsymbol{g}_t^T(I)]^T$. But the artificial noise and wireless channel noise will lead to inevitable perturbation on $\hat{\boldsymbol{g}}_t$, which yields a trade-off between accuracy and privacy.

8.3 Performance Analysis and System Optimization

8.3.1 Assumptions

For the tractability of theoretical analysis, we make the following assumptions.

Assumption 17 (Strong Convexity) *The global loss function $F(\boldsymbol{\theta})$ is strongly convex with parameter μ, i.e.,*

$$F(\boldsymbol{\theta}) \geq F(\boldsymbol{\theta}') + \nabla F(\boldsymbol{\theta}')^T(\boldsymbol{\theta} - \boldsymbol{\theta}') + \frac{\mu}{2}||\boldsymbol{\theta} - \boldsymbol{\theta}'||^2, \ \forall \boldsymbol{\theta}, \ \boldsymbol{\theta}'. \tag{8.12}$$

Assumption 18 (Smoothness) *The global loss function $F(\boldsymbol{\theta})$ is Lipschitz continuous with parameter L, i.e.,*

$$F(\boldsymbol{\theta}) \leq F(\boldsymbol{\theta}') + \nabla F(\boldsymbol{\theta}')^T(\boldsymbol{\theta} - \boldsymbol{\theta}') + \frac{L}{2}||\boldsymbol{\theta} - \boldsymbol{\theta}'||^2, \ \forall \boldsymbol{\theta}, \ \boldsymbol{\theta}',$$

which implies the following inequality:

$$||\nabla F(\boldsymbol{\theta}) - \nabla F(\boldsymbol{\theta}')|| \leq L||\boldsymbol{\theta} - \boldsymbol{\theta}'||, \ \forall \boldsymbol{\theta}, \ \boldsymbol{\theta}' \in \mathbb{R}^d.$$

Assumption 19 (Block Gradient Bound) *In any learning round t, for any training data sample $(\boldsymbol{x}, y)$, the sample-wise loss function is upper bounded by a given constant γ_t, i.e.,*

$$||\nabla f(\boldsymbol{x}, y; \boldsymbol{\theta}_t)|| \leq \gamma_t, \ \forall t. \tag{8.13}$$

Based on triangular inequality, for edge device k, there always exists a constant $\zeta_{k,t}(i) \leq \gamma_t$ satisfying $\left|\left|\boldsymbol{g}_{k,t}(i)\right|\right| \leq \zeta_{k,t}(i)$.

8.3.2 Convergence Analysis

To characterize the convergence behavior of the RIS-enabled FEEL system, we derive a bound on the average optimality gap, i.e., the expected value of the gap between the optimal value F^* and $\mathbb{E}[F(\boldsymbol{\theta}_{T+1}) - F^*]$, in the following theorem.

Theorem 8.1 *Under Assumptions 1 and 2, given learning rate $\lambda = 1/L$, after T learning iterations, the averaged optimality gap is upper bounded as*

$$\mathbb{E}\left[F\left(\boldsymbol{\theta}_{T+1}\right)\right] - F^* \leq \left(1 - \frac{\mu}{L}\right)^T \left[F\left(\boldsymbol{\theta}_1\right) - F^*\right] + \frac{1}{4L} \times \\ \frac{e}{(KD)^2} \sum_{t=1}^{T} \left(1 - \frac{\mu}{L}\right)^{T-t} \left(\sum_{k=1}^{K}\sum_{i=1}^{I} \sigma_{k,t}^2(i) + \sum_{i=1}^{I} \frac{N_0}{\eta_t(i)}\right). \tag{8.14}$$

Theorem 8.1 shows that the optimality gap $\mathbb{E}[F(\boldsymbol{\theta}_{T+1}) - F^*]$ is upper bounded by two terms. The first term approaches to 0 as T goes to infinity, while the second term does not diminish. This non-diminishing gap is related to $\boldsymbol{\Theta}_t(i)$ as the selection of $\eta_t(i)$ depends on $\boldsymbol{\Theta}_t(i)$, which will be shown in Theorem 8.3.

8.3.3 Privacy-Preserving Mechanism

To alleviate privacy concerns for edge devices, we next present a privacy-preserving scheme based on DP. In the t-th learning round, for edge device k, the disclosed signals regarding its local dataset $\mathcal{D}_k$ are $\{\boldsymbol{g}_{k,t}(i)\}$ across I blocks. We rewrite received signal $\boldsymbol{r}_t(i)$ as

$$\boldsymbol{r}_t(i) = \sqrt{\eta_t(i)} D_k \boldsymbol{g}_{k,t}(i) + \sqrt{\eta_t(i)} \sum_{j \neq k} D_j \boldsymbol{g}_{j,t}(i) + \boldsymbol{q}_t(i), \tag{8.15}$$

where

$$\boldsymbol{q}_t(i) = \sqrt{\eta_t(i)} \sum_{j=1}^{K} \boldsymbol{n}_{j,t}(i) + \boldsymbol{w}_t(i). \tag{8.16}$$

Denote $\boldsymbol{q}_t = [\boldsymbol{q}_t^T(1), \ldots, \boldsymbol{q}_t^T(I)]^T$ as the noise across I blocks. In (8.15), the first term is the disclosed signal concerning $\mathcal{D}_k$ and the third term $\boldsymbol{q}_t(i)$ is Gaussian noise which provides privacy protection for this system.

The privacy level (ϵ, δ) of each edge device depends on the sensitivity of the disclosed noise-free function with respect to its local dataset. The definition of l_2-sensitivity is listed as follows [3].

Definition 8.2 For an arbitrary mechanism $\mathcal{M}$ on the transmitted signals. The l_2-sensitivity Δ is defined as the maximum difference of the outputs from two neighboring datasets $\mathcal{D}_k$ and $\mathcal{D}'_k$, i.e.,

$$\Delta = \max_{\mathcal{D}_k, \mathcal{D}'_k} \left|\left|\mathcal{M}(\mathcal{D}_k) - \mathcal{M}\left(\mathcal{D}'_k\right)\right|\right|_2. \tag{8.17}$$

Only the first term in (8.15) has the potential to disclose information of $\mathcal{D}_k$. Denote $\boldsymbol{u}_{k,t}$ as the difference between the outputs from two neighboring datasets $\mathcal{D}_k$ and $\mathcal{D}'_k$, which is given by

$$\boldsymbol{u}_{k,t} = [h_{k,t}(1)\alpha_{k,t}(1)\Delta \boldsymbol{g}_{k,t}^T(1), \ldots, h_{k,t}(I)\alpha_{k,t}(I)\Delta \boldsymbol{g}_{k,t}^T(I)]^T, \tag{8.18}$$

where

$$\Delta \boldsymbol{g}_{k,t}(i) = \frac{1}{D} \sum_{(\boldsymbol{x},y) \in \mathcal{D}_k} \nabla f_i(\boldsymbol{x}, y; \boldsymbol{\theta}_t) - \frac{1}{D} \sum_{(\boldsymbol{x},y) \in \mathcal{D}'_k} \nabla f_i(\boldsymbol{x}, y; \boldsymbol{\theta}_t)$$

with $\nabla f_i(\boldsymbol{x}, y; \boldsymbol{\theta}_t)$ denoting the $(i-1)e+1$-th to ie-th elements in $\nabla f(\boldsymbol{x}, y; \boldsymbol{\theta}_t)$. Then the l_2-sensitivity $\Delta_{k,t}$ of device k is

$$\Delta_{k,t} = \max_{\mathcal{D}_k, \mathcal{D}'_k} \left|\left|\boldsymbol{u}_{k,t}\right|\right|_2. \tag{8.19}$$

By using Jensen inequality, the triangular inequality, and Assumption 3, we have

$$\Delta_{k,t} \leq 2\gamma_t \max_{i\in[I]} \sqrt{\eta_t(i)}. \tag{8.20}$$

Based on (8.20), the privacy constraint for all edge devices is given in the following theorem.

Theorem 8.2 *For any fixed sequence of* $\{\eta_t(i), \sigma_{k,t}(i)\}_{t=1}^T$, *the RIS-enabled FEEL system satisfies* (ϵ, δ)*-DP if we have*

$$\sum_{t=1}^{T} \frac{4\gamma_t^2}{\xi_t} \max_{i\in[I]} \{\eta_t(i)\} \leq \mathcal{R}_{dp}(\epsilon, \delta), \tag{8.21}$$

where

$$\xi_t \leq \eta_t(i) \sum_{k=1}^{K} \sigma_{k,t}^2(i) + N_0, \ \forall i, t, \tag{8.22}$$

denotes a lower bound of the power of the effective noise $\boldsymbol{q}_t(i)$, *and* $\mathcal{R}_{dp}(\epsilon, \delta) = \left(\sqrt{\epsilon + \left[C^{-1}(1/\delta)\right]^2} - C^{-1}(1/\delta)\right)^2$ *with* $C^{-1}(x)$ *denoting the inverse function of* $C(x) = \sqrt{\pi} x e^{x^2}$.

The proof of the above theorem directly follows [7]. Theorem 8.1 illustrates that both the artificial noise and the receiver noise contribute to the privacy protection. Besides, (8.22) shows that ξ_t is the lowest effective noise power across I communication blocks in the t-th learning round. Besides, the privacy guarantee depends the joint effect of ξ_t and the maximum transmit power scalar $\eta_t(i)$ among I communication blocks.

8.3.4 System Optimization

By the above analysis, it is evident that we can optimize the considered FEEL system by solving the following problem:

$$\underset{\{\eta_t, \xi_t, \boldsymbol{\Theta}_t, \sigma_{k,t}\}}{\text{minimize}} \quad \sum_{t=1}^{T} \left(1 - \frac{\mu}{L}\right)^{-t} \left(\sum_{k=1}^{K} \sum_{i=1}^{I} \sigma_{k,t}^2(i) + \sum_{i=1}^{I} \frac{N_0}{\eta_t(i)}\right) \tag{8.23a}$$

$$\text{subject to} \quad \sum_{t=1}^{T} \frac{4\gamma_t^2}{\xi_t} \max_{i\in[I]} \{\eta_t(i)\} \leq \mathcal{R}_{dp}(\epsilon, \delta), \tag{8.23b}$$

$$\xi_t \leq \eta_t(i) \sum_{k=1}^{K} \sigma_{k,t}^2(i) + N_0, \ \forall i, t, \tag{8.23c}$$

$$\frac{\eta_t(i)}{|\boldsymbol{m}_t^T(i)\boldsymbol{\Theta}_t(i)\boldsymbol{h}_{k,t}^r(i)+h_{k,t}^d(i)|^2}[D^2\zeta_{k,t}^2(i) + e\sigma_{k,t}^2(i)] \leq P_0, \ \forall i, k, t, \tag{8.23d}$$

$$|\boldsymbol{\Theta}_t(i)(n,n)| = 1, \quad \forall i, n, t, \tag{8.23e}$$

where (8.23b) and (8.23c) denote the privacy constraints for all edge devices as presented in (8.21) and (8.22), respectively, (8.23d) represents the power constraints by substituting $h_{k,t}(i) = \boldsymbol{m}_t^T(i)\boldsymbol{\Theta}_t(i)\boldsymbol{h}_{k,t}^r(i) + h_{k,t}^d(i)$ into (8.9), and (8.23e) indicates the unit modulus constraints of all reflecting elements at RIS with $\boldsymbol{\Theta}_t(i)(n,n)$ denoting as the (n,n)-th entry of $\boldsymbol{\Theta}_t(i)$.

8.4 A Two-Step Alternating Low-Rank Optimization Framework

This section provides a two-step alternating minimization framework for solving Problem (8.23).

8.4.1 Co-design of Artificial Noise and Power Scalar

Fix the phase-shift matrix $\boldsymbol{\Theta}_t(i)$, Problem (8.23) can be rewritten as follows:

$$\underset{\{\eta_t,\sigma_{k,t},\xi_t\}}{\text{minimize}} \quad \sum_{t=1}^{T}\left(1-\frac{\mu}{L}\right)^{-t}\left(\sum_{k=1}^{K}\sum_{i=1}^{I}\sigma_{k,t}^2(i)+\sum_{i=1}^{I}\frac{N_0}{\eta_t(i)}\right) \tag{8.24a}$$

$$\text{subject to} \quad \sum_{t=1}^{T}\frac{4\gamma_t^2}{\xi_t}\max_{i\in[I]}\{\eta_t(i)\} \leq \mathcal{R}_{dp}(\epsilon,\delta), \tag{8.24b}$$

$$\xi_t \leq \eta_t(i) \sum_{k=1}^{K} \sigma_{k,t}^2(i) + N_0, \ \forall i, t, \tag{8.24c}$$

$$\frac{D^2\zeta_{k,t}^2(i) + e\sigma_{k,t}^2(i)}{|h_{k,t}(i)|^2}\eta_t(i) \leq P_0, \ \forall i, k, t, \tag{8.24d}$$

where $h_{k,t}(i) = \boldsymbol{m}_t^T(i)\,\boldsymbol{\Theta}_t(i)\,\boldsymbol{h}_{k,t}^r(i) + h_{k,t}^d(i)$ denotes the composite channel. It is worth noting that Problem (8.24) can be transformed into a convex problem which can be tackled by KKT conditions. The details are summarized in Theorem 8.3. For notation ease, we denote τ_t as the number of communication blocks.

Theorem 8.3 *The optimal solution to Problem* (8.24) *is given by*

$$\xi_t = N_0,\ \forall t, \quad \sigma_{k,t}(i) = 0,\ \forall i, k, t, \tag{8.25}$$

and the selection of $\{\eta_t(i)\}$ *depends on the joint effect of privacy and power, which yields the following three cases:*

(a) If $\tau_t = I$ *for all learning rounds, then the optimal* $\{\eta_t(i)\}_{i=1}^I$ *remains a constant* η_t *with*

$$\eta_t = \frac{N_0\sqrt{I}}{2\gamma_t\beta^{\frac{1}{2}}}\left(1-\frac{\mu}{L}\right)^{-\frac{t}{2}} \wedge \eta_t < P_0\min_{i,k}\frac{|h_{k,t}(i)|^2}{D^2\zeta_{k,t}^2(i)}, \tag{8.26}$$

where β *is selected to strictly satisfy (8.24b), i.e.,*

$$\sum_{t=1}^{T}\frac{4\gamma_t^2}{N_0}\eta_t = \sum_{t=1}^{T}\frac{2\sqrt{I}\gamma_t}{\beta^{\frac{1}{2}}}\left(1-\frac{\mu}{L}\right)^{-\frac{t}{2}} = \mathcal{R}_{dp}(\epsilon,\delta). \tag{8.27}$$

(b) It is a generalized version of (a) with $\tau_t \in [I]$, *and the optimal* $\eta_t(i)$ *is given by*

$$\eta_t(i) = \min\left\{\frac{N_0\sqrt{\tau_t}}{2\gamma_t\beta^{\frac{1}{2}}}\left(1-\frac{\mu}{L}\right)^{-\frac{t}{2}},\ P_0\min_k\frac{|h_{k,t}(i)|^2}{D^2\zeta_{k,t}^2(i)}\right\}, \tag{8.28}$$

where β *and* τ_t *are selected to strictly satisfy (8.24b), i.e.,*

$$\sum_{t=1}^{T}\frac{4\gamma_t^2}{N_0}\max_{i\in[I]}\{\eta_t(i)\} = \sum_{t=1}^{T}\frac{4\gamma_t^2}{N_0}\min_{i\in[I]}\left\{\frac{N_0\sqrt{\tau_t}}{2\gamma_t\beta^{\frac{1}{2}}}\left(1-\frac{\mu}{L}\right)^{-\frac{t}{2}},\right.$$
$$\left. P_0\max_{i\in[I]}\left\{\min_{k\in[K]}\frac{|h_{k,t}(i)|^2}{D^2\zeta_{k,t}^2(i)}\right\}\right\} = \mathcal{R}_{dp}(\epsilon,\delta).$$

(c) If $\tau_t = 0$ *holds for all learning rounds, i.e.,*

$$\sum_{t=1}^{T}\frac{4P_0\gamma_t^2}{N_0}\max_{i\in[I]}\left\{\min_{k\in[K]}\frac{|h_{k,t}(i)|^2}{D^2\zeta_{k,t}^2(i)}\right\} < \mathcal{R}_{dp}(\epsilon,\delta), \tag{8.29}$$

the unique optimal power scalar $\eta_t(i)$ is given by

$$\eta_t(i) = P_0 \min_{k\in[K]} \frac{|h_{k,t}(i)|^2}{D^2\zeta_{k,t}^2(i)}. \tag{8.30}$$

From Theorem 8.3, the power of artificial noise becomes zero, which indicates that the additive channel noise in model aggregation serves as an inherent privacy-preserving mechanism to guarantee DP levels for each edge device [10].

We present the advantages achieved through the utilization of RIS in the aforementioned three scenarios. By leveraging the reconfigurable nature of RIS, it is observed that the FEEL system incorporating RIS is capable of establishing superior channel conditions compared to FEEL systems without RIS. Additionally, referring to Eqs. (8.28) and (8.30), the inclusion of RIS in the FEEL system enables higher transmit power and enhances the power of received signals at the edge server, thereby resulting in improved learning accuracy. However, in the case of stringent privacy requirements, denoted as Case (a), the learning accuracy remains constrained by privacy concerns and cannot be enhanced through the use of RIS.

8.4.2 *Design of Phase-Shift Matrix*

On the other hand, given the artificial noise and power scalar $\{\sigma_{k,t}(i), \eta_t(i)\}$, Problem (8.23) is reduced to the following feasibility detection problem:

$$\text{find} \quad \boldsymbol{\Theta}_t(i) \tag{8.31a}$$

$$\text{subject to} \quad \frac{\eta_t(i)\, D^2\zeta_{k,t}^2(i)}{|\boldsymbol{m}_t^T(i)\boldsymbol{\Theta}_t(i)\boldsymbol{h}_{k,t}^r(i)+h_{k,t}^d(i)|^2} \le P_0,\ \forall i,k,t, \tag{8.31b}$$

$$|\boldsymbol{\Theta}_t(i)(n,n)| = 1,\ \forall i,n,t. \tag{8.31c}$$

By denoting $v_{n,t}(i) = e^{j\phi_{n,t}(i)}$ and $\boldsymbol{c}_{k,t}^H(i) = \boldsymbol{m}_t^T(i)\,\mathrm{diag}(\boldsymbol{h}_{k,t}^r(i))$, we have $\boldsymbol{m}_t^T(i)\,\boldsymbol{\Theta}_t(i)\,\boldsymbol{h}_{k,t}^r(i) = \boldsymbol{c}_{k,t}^H(i)\,\boldsymbol{v}_t(i)$, where $\boldsymbol{v}_t(i) = [v_{1,t}(i), \ldots, v_{N,t}(i)]^T$. We then transform Problem (8.31) into the following problem:

$$\text{find} \quad \boldsymbol{v}_t(i) \tag{8.32a}$$

$$\text{subject to} \quad \frac{\eta_t(i)D^2\zeta_{k,t}^2(i)}{|\boldsymbol{c}_{k,t}^H(i)\boldsymbol{v}_t(i) + h_{k,t}^d(i)|^2} \le P_0,\ \forall k,t, \tag{8.32b}$$

$$|v_{n,t}(i)| = 1,\ \forall n \in [N]. \tag{8.32c}$$

Nevertheless, Problem (8.32) is nonconvex. By introducing an auxiliary variable $\iota_t(i)$, problem (8.32) can be equivalently reformulated as a homogeneous quadratically constrained quadratic programming (QCQP) problem as follows:

$$\text{find} \quad \hat{\boldsymbol{v}}_t(i) \tag{8.33a}$$

$$\text{subject to} \quad \hat{\boldsymbol{v}}_t^H(i)\boldsymbol{R}_{k,t}(i)\hat{\boldsymbol{v}}_t(i) + |h_{k,t}^d(i)|^2$$

$$\geq \eta_t(i)D^2\zeta_{k,t}^2(i)/P_0, \ \forall i, k, t, \tag{8.33b}$$

$$|\hat{v}_{n,t}(i)| = 1, \ \forall n \in [N+1], \tag{8.33c}$$

where

$$\boldsymbol{R}_{k,t}(i) = \begin{bmatrix} \boldsymbol{c}_{k,t}(i)\left(\boldsymbol{c}_{k,t}(i)\right)^H & \boldsymbol{c}_{k,t}(i)h_{k,t}^d(i) \\ \left(\boldsymbol{c}_{k,t}(i)h_{k,t}^d(i)\right)^H & 0 \end{bmatrix}, \ \hat{\boldsymbol{v}}_t(i) = \begin{bmatrix} \boldsymbol{v}_t(i) \\ \iota_t(i) \end{bmatrix}.$$

Let $\hat{\boldsymbol{v}}_t^*(i) = [\widetilde{\boldsymbol{v}}_t^*(i)^T, \iota_t^*(i)]^T$ denote a feasible $\hat{\boldsymbol{v}}_t(i)$ to problem (8.33). Then a feasible solution $\boldsymbol{v}_t^*(i)$ to problem (8.32) can be immediately recovered by setting $\boldsymbol{v}_t^*(i) = \widetilde{\boldsymbol{v}}_t^*(i)/\iota_t^*(i)$. A feasible solution $\boldsymbol{\Theta}_t^*(i)$ to problem (8.31) can thus be expressed as $\boldsymbol{\Theta}_t^*(i) = \text{diag}\,(\boldsymbol{v}_t^*(i))$.

Since $\hat{\boldsymbol{v}}_t^H(i)\,\boldsymbol{R}_{k,t}(i)\,\hat{\boldsymbol{v}}_t(i) = \text{Tr}(\boldsymbol{R}_{k,t}(i)\,\hat{\boldsymbol{v}}_t(i)\,\hat{\boldsymbol{v}}_t^H(i))$, we denote $\boldsymbol{V}_t(i)$ as the lifting matrix of $\hat{\boldsymbol{v}}_t(i)$, where $\boldsymbol{V}_t(i) = \hat{\boldsymbol{v}}_t(i)\,\hat{\boldsymbol{v}}_t^H(i)$ and $\text{rank}\,(\boldsymbol{V}_t(i)) = 1$. Problem (8.32) is equivalent to the following problem:

$$\text{find} \quad \boldsymbol{V}_t(i) \tag{8.34a}$$

$$\text{subject to} \quad \text{Tr}\left(\boldsymbol{R}_{k,t}(i)\boldsymbol{V}_t(i)\right) + |h_{k,t}^d(i)|^2$$

$$\geq \eta_t(i)D^2\zeta_{k,t}^2(i)/P_0, \ \forall i, k, t, \tag{8.34b}$$

$$\boldsymbol{V}_t(i)(n,n) = 1, \ \forall n \in [N+1], \tag{8.34c}$$

$$\text{rank}\,(\boldsymbol{V}_t(i)) = 1, \ \forall i, t, \tag{8.34d}$$

$$\boldsymbol{V}_t(i) \succeq \boldsymbol{0}, \ \forall i, t, \tag{8.34e}$$

where $\boldsymbol{V}_t(i)(n,n)$ denotes the (n,n)-entry of matrix $\boldsymbol{V}_t(i)$.

The rank one constraint hinders the solving of this problem. To circumvent it, we resort to SDR technique [12] by simply dropping the rank-one constraints and then focus on the simplified convex SDP problem which can be efficiently tackled by existing convex optimization solvers such as CVX [6].

Without causing confusion, we omit parameter i for notational ease, e.g., $\boldsymbol{\Theta}_t(i)$ as $\boldsymbol{\Theta}_t$. We present the two-step iterative framework in Algorithm 1 for solving problem $\mathscr{P}$.

Algorithm 11: Two-step alternating minimization for solving Problem (8.23) in the RIS-enabled FEEL system

Input: The phase-shift matrix $\boldsymbol{\Theta}_0$, the privacy level (ϵ, δ), and the maximum iteration number J.

Initialize $j \leftarrow 1$ to denote the number of iterations.
repeat
  Given $\boldsymbol{\Theta}_{j-1}$, obtain solution $(\eta_j, \sigma_{k,j})$ by solving Problem (8.24).
  Given $(\eta_j, \sigma_{k,j})$, obtain solution $\boldsymbol{\Theta}_j$ by solving problem Problem (8.31).
  Update $j \leftarrow j + 1$.
until *(8.23a) is below a given threshold or* $j > J$;

Output: $\eta_t(i) \leftarrow \eta_j$, $\boldsymbol{\Theta}_t(i) \leftarrow \boldsymbol{\Theta}_j$, $\sigma_{k,t}(i) \leftarrow \sigma_{k,j}$.

8.5 Simulation Results

8.5.1 Simulation Settings

The performance of the presented scheme is demonstrated on ridge regression, with the sample-wise loss function given as

$$f(\boldsymbol{x}, y; \boldsymbol{\theta}) = \frac{1}{2}\left|\left|\boldsymbol{\theta}^T\boldsymbol{x} - y\right|\right|^2 + \upsilon||\boldsymbol{\theta}||^2, \tag{8.35}$$

where the penalty coefficient is set as $\upsilon = 5 \times 10^{-5}$. Randomly generating a dataset of scale $|\mathcal{D}| = 10^4$ and set the model dimension d to be 10. To be specific, the training samples $\boldsymbol{x}$ are drawn i.i.d. according to $\mathcal{N}(\mathbf{0}, \boldsymbol{I}_d)$ while the corresponding true label y is given by $y = \boldsymbol{x}(2) + 3\boldsymbol{x}(5) + 0.2z_0$, where the data noise z_0 is drawn i.i.d. from $\mathcal{N}(0, 1)$. We uniformly divide the total dataset $\mathcal{D}$ into K local datasets.

We assume that the channels suffer from Rice fading [17], and the channel coefficients are given by

$$\boldsymbol{\varrho} = \sqrt{\frac{\kappa}{1+\kappa}}\boldsymbol{\varrho}_{LoS} + \sqrt{\frac{1}{1+\kappa}}\boldsymbol{\varrho}_{NLoS}, \tag{8.36}$$

where κ represents the Rician factor, $\boldsymbol{\varrho}_{LoS}$ denotes the deterministic line-of-sight (LoS) component, and $\boldsymbol{\varrho}_{NLoS}$ denotes the non-line-of-sight (NLoS) component. In our simulation, we set $\boldsymbol{\varrho}_{LoS} = 1$ and generate $\boldsymbol{\varrho}_{NLoS}$ by autoregressive (AR) scheme, which is given by

$$\boldsymbol{\varrho}_{NLoS}(i) = \rho\boldsymbol{\varrho}_{NLoS}(i-1) + \sqrt{1-\rho^2}\boldsymbol{\vartheta}(i), \tag{8.37}$$

where ρ denotes the correlation coefficient which is set to be 1 and $\boldsymbol{\vartheta}(i)$ is drawn based on an innovation process satisfying $\boldsymbol{\vartheta}(i) \sim \mathcal{N}(\mathbf{0}, \boldsymbol{I})$. Hence, the channel coefficients in the i-th communication block of learning round t are given by

Table 8.1 System parameters

Parameter	Value	Parameter	Value	Parameter	Value
K	10	N	30	I	5
D_k	1000	r	0.1	T	30
κ_{IB}	5	κ_{DI}	0	κ_{DB}	5
ϵ	20	δ	0.01	SNR	35

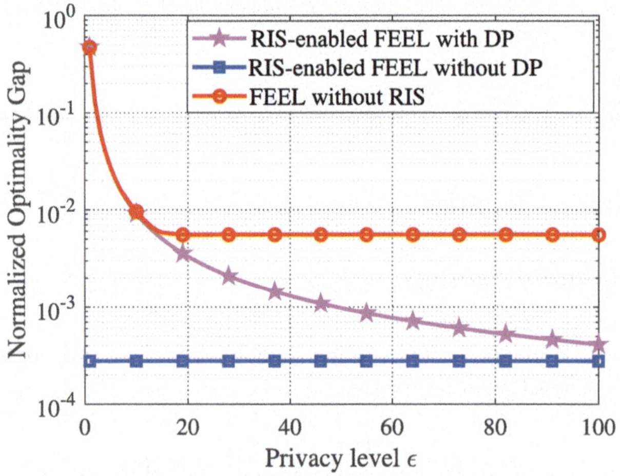

Fig. 8.2 Learning accuracy of three RIS schemes under different privacy levels

$\boldsymbol{m}_t(i) = \boldsymbol{\varrho}_{IB}$, $\boldsymbol{h}^r_{k,t}(i) = \boldsymbol{\varrho}_{DI}$, and $h^d_{k,t}(i) = \varrho_{DB}$, whose Rician factors are given by κ_{IB}, κ_{DI}, and κ_{DB}, respectively. Specific parameters setting are summarized in Table 8.1.

8.5.2 *Learning Performance*

In Fig. 8.2, we present the correlation between learning accuracy and privacy level ϵ. The results reveal that as the privacy level is relaxed, indicated by larger values of ϵ, the RIS-enabled FEEL system exhibits a significant improvement in learning performance compared to the scenario without RIS. This improvement can be attributed to the mitigation of the adverse effects of unfavorable wireless channel propagations. However, it is important to note that when the privacy level is stringent, as stated in Theorem 8.3, the presence of RIS does not impact the learning performance. Furthermore, when comparing the learning accuracy of the RIS-enabled FEEL system with DP and the case without DP, we observe that achieving privacy guarantees comes at the cost of reduced learning accuracy. In

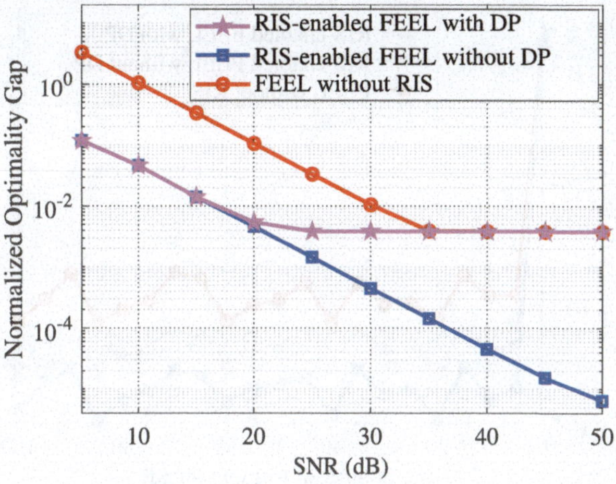

Fig. 8.3 Learning accuracy of three RIS schemes under various noisy environments

other words, higher privacy levels (smaller values of ϵ) correspond to lower learning accuracy.

In Fig. 8.3, we examine the impact of varying signal-to-noise ratio (SNR) levels on the learning accuracy of RIS-enabled FEEL systems operating in a noisy wireless environment. Interestingly, we observe that as the SNR increases, the DP-restricted schemes consistently achieve a similar learning accuracy of approximately 4×10^{-3}, while adhering to the same privacy and power constraints. This finding highlights the robustness of these schemes in maintaining learning performance despite changes in SNR. Moreover, it is worth noting that the presence of RIS in FEEL systems leads to a significant enhancement in learning accuracy, particularly in scenarios with low SNR. This suggests that RIS-enabled FEEL systems exhibit greater adaptability in challenging wireless channels affected by noise. Additionally, Fig. 8.3 demonstrates that when the SNR is below 20, the RIS-enabled FEEL system with DP achieves comparable accuracy to the case without DP. This implies that privacy guarantees can be achieved without compromising learning accuracy, aligning with the concept of privacy being freely ensured [10].

Figure 8.4 illustrates the impact of the number of learning rounds T or total communication blocks on the learning accuracy, taking into account the influence of wireless noise introduced in each communication block. Interestingly, when considering the effect of noise, we observe that the number of learning rounds T does affect the accuracy, even after running a sufficient number of training rounds for gradient descent. This effect is attributed to the adaptive selection of $\{\beta, \tau_t\}$ as outlined in **Theorem 3**. Specifically, the flexibility of β allows for the adjustment of $\eta_t(i)$ selection to satisfy all constraints as T increases. Furthermore, the RIS-enabled FEEL systems not only achieve higher accuracy but also exhibit improved robustness against the newly introduced wireless channel noise.

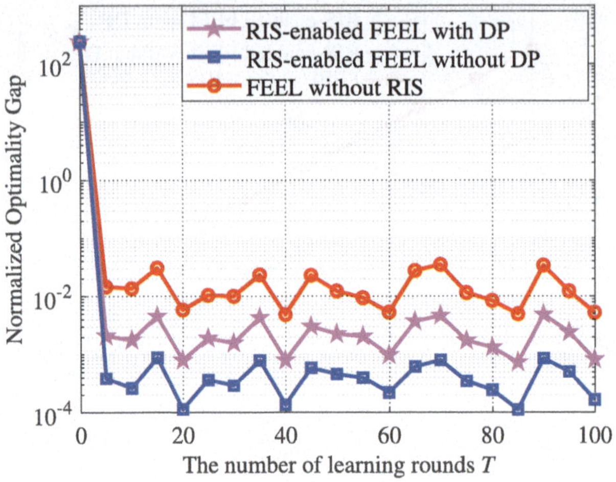

Fig. 8.4 Learning accuracy of three RIS schemes under different numbers of learning rounds

8.6 Summary

This chapter explored the potential of RIS and AirComp in enhancing the privacy and accuracy of FEEL. We analyzed the convergence behavior of the over-the-air FEEL algorithm and formulating a system optimization problem to achieve improved learning accuracy while adhering to privacy and transmit power constraints. To minimize the learning optimality gap, we introduced a two-step low-rank optimization framework that jointly optimized power allocation, artificial noise, and reflecting coefficients of RIS during the learning process. The proposed approach not only enhances the system SNR and receive signal power but also significantly improves learning accuracy while meeting privacy requirements.

References

1. N. Agarwal, A. T. Suresh, F. Yu, S. Kumar, and H. B. Mcmahan. cpSGD: Communication-efficient and differentially-private distributed SGD. *Adv. Neural Inf. Process. Syst.*, pages 7564–7575, 2018.
2. M. M. Amiri and D. Gündüz. Machine learning at the wireless edge: Distributed stochastic gradient descent over-the-air. *IEEE Trans. on Signal Process.*, 68:2155–2169, 2020.
3. C. Dwork, A. Roth, et al. The algorithmic foundations of differential privacy. *Found. Trends Theor. Comput. Sci.*, 9(3–4):211–407, 2014.
4. J. Geiping, H. Bauermeister, H. Dröge, and M. Moeller. Inverting gradients–how easy is it to break privacy in federated learning? *arXiv:2003.14053. [Online].* https://arxiv.org/pdf/2003.14053.pdf, 2020.

5. A. M. Girgis, D. Data, S. Diggavi, P. Kairouz, and A. T. Suresh. Shuffled model of federated learning: Privacy, communication and accuracy trade-offs. *IEEE J. Sel. Areas Commun. Inf. Theory (JSAIT)*, 2(1):464–478, 2021.
6. M. Grant and S. Boyd. CVX: Matlab software for disciplined convex programming, version 2.1. http://cvxr.com/cvx, Mar. 2014.
7. B. Hasırcıoğlu and D. Gündüz. Private wireless federated learning with anonymous over-the-air computation. In *2021 IEEE Int. Conf. Acoust. Speech Signal Process. (ICASSP)*, pages 5195–5199. IEEE, 2021.
8. B. Hitaj, G. Ateniese, and F. Perez-Cruz. Deep models under the GAN: information leakage from collaborative deep learning. In *Proc. of the 2017 ACM SIGSAC Conf. Comput. Commun. Secur.*, pages 603–618, 2017.
9. L. Li, L. Yang, X. Guo, Y. Shi, H. Wang, W. Chen, and K. B. Letaief. Delay analysis of wireless federated learning based on saddle point approximation and large deviation theory. *arXiv:2103.16994. [Online]*. https://arxiv.org/pdf/2103.16994.pdf, 2021.
10. D. Liu and O. Simeone. Privacy for free: Wireless federated learning via uncoded transmission with adaptive power control. *IEEE J. Sel. Areas Commun.*, 39(1):170–185, 2020.
11. H. Liu, X. Yuan, and Y.-J. A. Zhang. Reconfigurable intelligent surface enabled federated learning: A unified communication-learning design approach. *IEEE Trans. Wireless Commun.*, 2021. doi:10.1109/TWC.2021.3086116.
12. Z.-Q. Luo, W.-K. Ma, A. M.-C. So, Y. Ye, and S. Zhang. Semidefinite relaxation of quadratic optimization problems. *IEEE Signal Process. Mag.*, 27(3):20–34, 2010.
13. M. Seif, R. Tandon, and M. Li. Wireless federated learning with local differential privacy. In *2020 IEEE Int. Symp. on Inf. Theory (ISIT)*, pages 2604–2609. IEEE, 2020.
14. T. Sery and K. Cohen. On analog gradient descent learning over multiple access fading channels. *IEEE Trans. Signal Process.*, 68:2897–2911, 2020.
15. Z. Wang, J. Qiu, Y. Zhou, Y. Shi, L. Fu, W. Chen, and K. B. Letaief. Federated learning via intelligent reflecting surface. *IEEE Trans. Wireless Commun.*, 2021.
16. N. Wu, F. Farokhi, D. Smith, and M. A. Kaafar. The value of collaboration in convex machine learning with differential privacy. In *2020 IEEE Secur. Priv. (SSP)*, pages 304–317. IEEE, 2020.
17. Q. Wu and R. Zhang. Intelligent reflecting surface enhanced wireless network via joint active and passive beamforming. *IEEE Trans. Wireless Commun.*, 18(11):5394–5409, 2019.
18. Y. Yang, Y. Zhou, Y. Wu, and Y. Shi. Differentially private federated learning via reconfigurable intelligent surface. *IEEE Internet of Things Journal*, 9(20):19728–19743, 2022.
19. G. Zhu, Y. Wang, and K. Huang. Broadband analog aggregation for low-latency federated edge learning. *IEEE Trans. Wireless Commun.*, 19(1):491–506, 2019.
20. L. Zhu and S. Han. Deep leakage from gradients. In *Federated Learning*, pages 17–31. Springer, 2020.

Chapter 9
Trustworthy Federated Edge Learning via Blockchain

9.1 Overview

FEEL ensures privacy protection for edge devices, yet it faces vulnerabilities to attacks from malicious entities, such as model poisoning or adversarial attacks, as noted by So et al. [1]. To mitigate these risks, a variety of secure model aggregation algorithms have been devised, including geometric median, trimmed mean, and multi-Krum [2]. Despite the efficiency of the server-client architecture in FEEL, it is prone to server malfunctions, potentially resulting in single points of failure or tampering with the global model, thereby compromising the entire FEEL training process. To counter these challenges, the integration of blockchain technology has led to the development of decentralized B-FEEL [3–5]. This network leverages multiple servers for global model aggregation and employs various consensus protocols, such as Proof of Work (PoW), Proof of Stake (PoS), Raft, and Practical Byzantine Fault Tolerance (PBFT), to ensure the integrity of the global model [6, 7]. Consequently, B-FEEL increases system resilience against failures or attacks on servers and devices, establishing a trustworthy mechanism for global model aggregation through secure, blockchain-based consensus across multiple servers [8–11].

On the other hand, latency in the FEEL training process is a crucial metric in edge AI, particularly for improving communication efficiency in light of the fading nature of wireless channels [12–15]. In FEEL over wireless networks, edge devices transmit their local models while the edge server disseminates the shared global model via wireless links. This process involves computation latency, encompassing local training, global model aggregation, and block validation, and communication latency, including uplink and downlink transmissions for model updates and blockchain consensus protocol. These latencies are influenced by factors such as computational capabilities, dataset and model sizes, wireless communication techniques, bandwidth, and power budgets. Recent theoretical analysis and resource allocation studies have addressed FEEL latency [16–19]. However, in

Y. Zhou et al., *Federated Edge Learning*, Wireless Networks,
https://doi.org/10.1007/978-3-031-96649-1_9

wireless B-FEEL, latency becomes more complex due to additional rounds of cross-validation among edge servers for blockchain consensus. Therefore, characterizing and optimizing the latency of the wireless B-FEEL system is essential for reducing overall learning latency and enhancing system efficiency.

In this chapter, our objective is to design a B-FEEL that guarantees trustworthiness while minimizing latency, utilizing the consensus protocol of blockchain. We strategically employ the PBFT consensus protocol [20, 21], prioritizing high effectiveness and low energy consumption, albeit with some security trade-offs compared to other consensus protocols like PoW and Raft [7]. The latency in the B-FEEL system is dissected into computation and communication components, encompassing local training, global model aggregation, consensus protocol, and model dissemination. To optimize the system, we aim to minimize the long-term average learning latency by allocating bandwidth and power resources strategically. Subsequently, we transforming the long-term resource allocation problem into a MDP. For this MDP, we develop a DRL-based algorithm that ensures efficient and adaptive resource allocation with lower computational complexity from a long-term perspective. The developed DRL-based algorithm leverages a DNN to directly map current network information to optimal resource allocation decisions, thereby significantly reducing the complexity of the decision-making process.

9.2 System Model

9.2.1 Architecture of Wireless Blockchain-Based FEEL

As illustrated in Fig. 9.1, we consider a wireless B-FEEL system consisting of M edge servers and K edge devices. Each edge server, indexed by $\mathcal{B} = \{B_1, \cdots, B_m, \cdots, B_M\}$, is equipped with a mobile edge computing (MEC) server for executing computing tasks and is endowed with sufficient computation and storage resources for global model aggregation and the consensus protocol [3]. Concurrently, the K edge devices are indexed by $\mathcal{D} = \{D_1, \cdots, D_k, \cdots, D_K\}$, each with a local dataset denoted as $\mathcal{S} = \{\mathcal{S}_1, \cdots, \mathcal{S}_k, \cdots, \mathcal{S}_K\}$, and possess adequate computing and storage resources to perform local training in the B-FEEL process. Communication between edge servers occurs over wireless channels [3, 22].

Conventional server-client FEEL architectures are vulnerable to edge server malfunctions, including single points of failure and global model poisoning [23]. Blockchain technology offers a solution in wireless FEEL by introducing decentralization to counter these vulnerabilities [11, 24]. B-FEEL uses multiple edge servers and devices [25] to create a blockchain that secures blocks of data with cryptography and a consensus protocol. This approach ensures data accuracy and immutability, providing a resilient, transparent model aggregation platform that mitigates edge server failures and tampering risks."

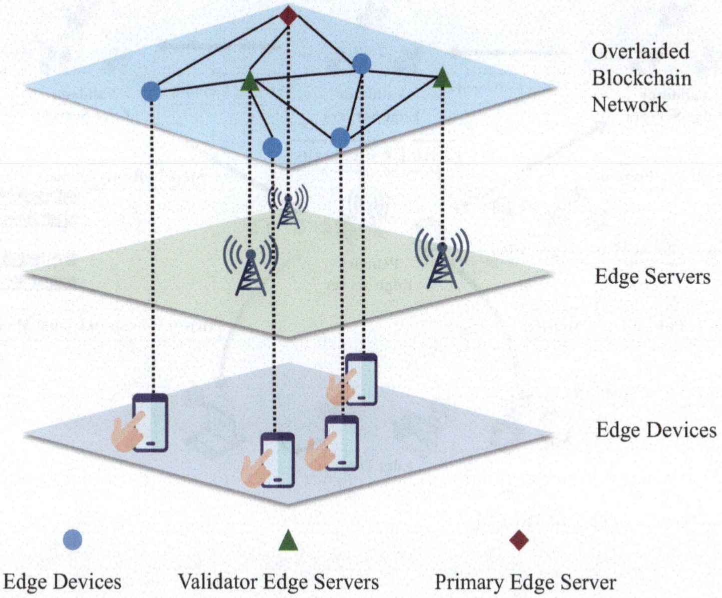

Fig. 9.1 Wireless blockchain-based Federated edge learning system

The B-FEEL system includes a primary edge server and multiple validator edge servers for block production and consensus. In each training round, edge devices train their local models and upload them to the primary server. This server verifies the local models and aggregates them into a global model using an immutable and trackable smart contract, primarily for secure global model aggregation. The local and global models are then compiled into a new block and broadcasted to the validator servers for verifying the global model's correctness. Once consensus is reached, the block is added to the blockchain, ensuring transparency and traceability. The validated global model is subsequently disseminated to edge devices. Thus, the consensus protocol plays a crucial role in maintaining the uniformity and security of this decentralized system.

9.2.2 Procedures of B-FEEL

In this chapter, we utilize the PBFT consensus protocol within our B-FEEL system for global model aggregation, a more effective approach than PoW. The training procedures under PBFT in the wireless B-FEEL system are detailed in Fig. 9.2.

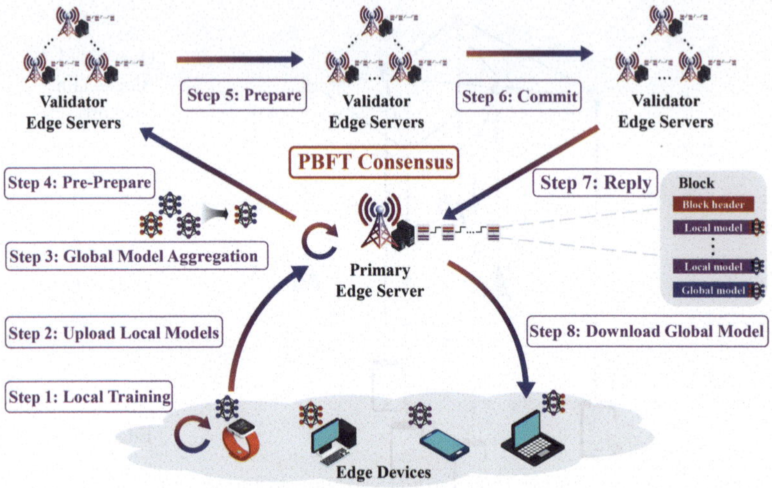

Fig. 9.2 The procedures of B-FEEL

9.2.2.1 Local Training

The local model parameter can be updated usint the stochastic gradient descent algorithm as follows

$$\boldsymbol{w}_k^t = \boldsymbol{w}_k^{t-1} - \eta \nabla F_k(\boldsymbol{w}_k^{t-1}; \mathcal{S}_k), \tag{9.1}$$

where F_k denotes the local loss function, calculated on a batch of samples $\mathcal{S}_k$ randomly selected from the local dataset, $\boldsymbol{w}_k^t$ represents the model of edge device D_k in round t, $\nabla F_k(\boldsymbol{w}_k^{t-1}; \mathcal{S}_k)$ denotes the gradient of $F_k(\cdot)$, and η is the learning rate. In this step, the computation latency is the local model parameter calculation at the edge devices.

9.2.2.2 Upload Local Models

Using allocated wireless resources, each edge device $D_k \in \mathcal{D}$ uploads its local model to the primary edge server in the form of transactions $\langle \boldsymbol{w}_k, D_k \rangle$. Here, $\langle \boldsymbol{w}_k, D_k \rangle$ represents a data packet that includes the local model parameter $\boldsymbol{w}_k$ and is digitally signed by the edge device D_k to ensure authenticity and integrity. Digital signatures help validate data packets, preventing tampering by unauthorized entities. In the context of blockchain, a transaction encompasses various activities like cryptocurrency exchanges, contracts, and records, documenting behaviors within the blockchain. Notably, a primary edge server is appointed prior to each training round, with the potential for rotation among all available edge servers. This arrangement ensures that all edge devices and servers are apprised of the identity

Algorithm 12: Pseudocode of B-FEEL

1. **Require:** Edge servers set $\mathcal{B}$, edge devices set $\mathcal{D}$, the dataset of edge devices $\mathcal{S}$.
2. **Ensure:** The global model $\boldsymbol{w}_g$ and a blockchain that contains training information.
3. Initialize the blockchain $\boldsymbol{B}$ and global model $\boldsymbol{w}_g^0$.
4. For each round $t = 1, 2, \ldots$:
   a. Rotate primary edge server among edge servers.
   b. Allocate bandwidth and power resources.
   c. For each device $D_k \in \mathcal{D}$:
      i. D_k executes local training on its local data $\mathcal{S}_k$ to generate local model $\boldsymbol{w}_k^t$.
      ii. D_k uploads its local model to primary edge server.
   d. Primary edge server validates the local models $\{\boldsymbol{w}_k^t\}_{D_k \in \mathcal{D}}$ and aggregates them to a global model $\boldsymbol{w}_g^t = \text{multi\_KRUM}(\{\boldsymbol{w}_k^t\}_{D_k \in \mathcal{D}})$.
   e. Primary edge server packs local models and global model to a new block and broadcasts the block to the validator edge servers.
   f. Primary edge server and validator edge servers reach consensus by PBFT protocol consensus (including pre-prepare, prepare, commit, reply steps) to guarantee the correctness of the global model and add the verified block to the blockchain.
   g. Primary edge server broadcasts global model $\boldsymbol{w}_g^t$ to edge devices.

function multi_KRUM ($\boldsymbol{w}^t_{k\, D_k \in \mathcal{D}}$)

1. Compute the Euclidean distances $s(D_k) = \sum_{D_j \neq D_k} \|\boldsymbol{w}_k - \boldsymbol{w}_j\|^2, \forall D_k \in \mathcal{D}$, where the sum runs over the $K - f - 2$ closest vectors and f is a specified parameter with respect to Byzantine tolerance.
2. Select the $K - f$ lowest distances and corresponding local models as $\hat{\mathcal{D}} \subset \mathcal{D}$.
3. Compute $\boldsymbol{w}_g = \frac{1}{|\hat{\mathcal{D}}|} \sum_{D_k \in \hat{\mathcal{D}}} \boldsymbol{w}_k$ by averaging.
4. **return** $\boldsymbol{w}_g$.

of the primary edge server at the commencement of each round. In this context, the primary source of communication latency stems from the duration required for uploading transactions from all edge devices.

9.2.2.3 Global Model Aggregation

After receiving data packets $\langle \boldsymbol{w}_k, D_k \rangle$ from all edge devices $D_k \in \mathcal{D}$, the primary edge server authenticates each transaction to verify the ownership of local models and then employs a smart contract for their aggregation into a global model. We apply the multi-KRUM algorithm for Byzantine-resilient secure global model aggregation [2, 26, 27], represented as:

$$\boldsymbol{w}_g^t = \text{multi\_KRUM}(\{\boldsymbol{w}_k^t, \forall D_k \in \mathcal{D}\}), \quad (9.2)$$

where $\boldsymbol{w}_g^t$ is the global model at round t, and $\boldsymbol{w}_k^t$ is the local model from edge device D_k. The function $\mathsf{multi\_KRUM}(\cdot)$ refers to the multi-KRUM algorithm, detailed in Algorithm 12. Subsequently, the local and global models are compiled into a new block, denoted as $B = \left\langle \{\langle \boldsymbol{w}_1, D_1 \rangle\}_{D_k \in \mathcal{D}}, \left\langle \boldsymbol{w}_g, B_p \right\rangle \right\rangle$, with B_p representing the primary edge server. In this phase, the computation latency involves transaction validation and model aggregation on the primary edge server.

9.2.2.4 Pre-prepare

Once the primary edge server generates a new block, it broadcasts this block along with a pre-prepare message, formatted as

$$\left\langle \text{PRE-PREPARE}, H_B, D(B), B, B_p \right\rangle .$$

H_B represents the block height, indicating the number of valid blocks in the blockchain, while $D(B)$ is a digest of the block B, such as its hash value. The primary edge server B_p signs this pre-prepare message for authentication. Validator edge servers, which include all edge servers except the primary one, then receive the new block. Their task is to verify the digital signature of the transactions and the new block, ensuring the validity of the new block's digital signature. Furthermore, they undertake the recalculation of the global model to authenticate the accuracy of the computations performed by the primary edge server. Upon successful validation, each validator edge server proceeds to store the newly formed block locally. In this intricate process, the computation latency is primarily consumed by the verification of the block's digital signature and the re-execution of the global model calculation. Concurrently, the communication latency arises from the dissemination of the block from the primary edge server to the validators.

9.2.2.5 Prepare

After their verification, prepare messages $\langle \text{PREPARE}, H_B, D(B), B_m \rangle$ are send out by the validator edge servers to the others, where B_m represents each validator server except for the primary server B_p. These messages serve as a confirmation that a substantial majority of validators have concurred with the addition of the new block. In this phase, the computation latency primarily involves the verification of the digital signatures embedded within the prepare messages by all participating edge servers. Meanwhile, the communication latency is incurred by the dissemination of these messages from the validator servers to the respective recipients.

9.2.2.6 Commit

Once a consensus is reached on the new block, the validator edge servers broadcast commit messages $\langle \text{COMMIT}, H_B, D(B), B_m \rangle$, applicable to all $B_m \in \mathcal{B}$. These messages are sent to all servers to confirm that the majority have agreed on the block. The computation latency in this phase involves validating the digital signatures of the commit messages at all edge servers. The communication latency is associated with the transmission of these messages from all the edge servers.

9.2.2.7 Reply

Once consensus is achieved, the new block is validated. Subsequently, all validator edge servers send reply messages $\langle \text{REPLY}, H_B, D(B), B_m \rangle$ back to the primary edge server, where $B_m \in \mathcal{B}$ and $B_m \neq B_p$. These messages indicate that the validators have stored the new block in the blockchain and are prepared for the next round of consensus. At this stage, the primary edge server's computation latency involves validating the digital signatures of these reply messages. The communication latency is due to the transmission of these messages from the validator servers to the primary server.

9.2.2.8 Download Global Model

In this final step, the primary edge server dispatches the validated global model back to the edge devices, facilitating the commencement of the next training round. Here, the communication latency is primarily attributed to the downloading of the global model by the edge devices.

9.2.3 *Communication Model*

The channel gain between edge device D_k and edge server B_m for a time-slot s is denoted by $h_{D_k,B_m}[s]$. Given that time-slots are typically short (e.g., 10 ms in the LTE protocol), but a B-FEEL training round spans over several seconds due to the complexities in model training and consensus protocol, each B-FEEL training round encompasses multiple time-slots [16, 18]. Considering our focus on minimizing total latency over numerous rounds, the variability of channel gains across multiple time-slots within a single round is deemed negligible. Therefore, we simplify our analysis by using the average channel gain over one round, represented as $h^t_{D_k,B_m} = \frac{1}{S}\sum_{s=1}^{S} h_{D_k,B_m}[tS+s]$, where S is the total number of time-slots in one round and t denotes the training round [18].

Thus, the achievable transmission rate from edge device D_k to edge server B_m can be denoted as

$$R^t_{D_k,B_m} = b^t_{D_k} \log_2 \left(1 + \frac{h^t_{D_k,B_m} p^t_{D_k}}{b^t_{D_k} N_0}\right), \tag{9.3}$$

where $b^t_{D_k}$ denotes bandwidth assigned to edge device D_k, and $h^t_{D_k,B_m}$ denotes the channel gain from edge device D_k to edge server B_m. In addition, $p^t_{D_k}$ and N_0 are the transmit power in edge device D_k and the power spectral density of additive white Gaussian noise (AWGN). In this chapter, we denote the maximum system bandwidth as $b^{\max}$. This is expressed as $\sum_{d \in \mathcal{D} \cup \mathcal{B}} b^t_d \leq b^{\max}$, indicating that the edge servers and edge devices collectively share the same broadband communication bandwidth [3]. The latency for transmitting a data packet of size ϖ from device D_k to edge server B_m thus can be calculated as

$$T_{D_k,B_m} = \frac{\varpi}{R^t_{D_k,B_m}}. \tag{9.4}$$

9.3 System Optimization

9.3.1 Latency Analysis for Wireless B-FEEL

In the subsequent analysis, we'll detail the steps and associated latencies for each phase in the t-th round of our wireless B-FEEL system. This system employs a synchronized transmission scheme to ensure all edge servers and devices progress in unison. Compared to a conventional wireless FEEL system without blockchain, the wireless B-FEEL system with blockchain integration incurs additional computational overhead, mainly due to the need for validating digital signatures of local models and reaching consensus on global model aggregation. We assume that generating or verifying one digital signature consumes ρ CPU cycles.

9.3.1.1 Local Training

The number of CPU cycles required for devices to execute the backpropagation algorithm with one sample is denoted by δ. Consequently, the latency for calculating the local model weights can be expressed as"

$$T^{\mathrm{cmp}}_{\mathrm{train}} = \max_{D_k \in \mathcal{D}} \left\{ \frac{s_{D_k} \delta}{f_{D_k}} \right\}, \tag{9.5}$$

where s_{D_k} denotes the batch size of edge device D_k and f_{D_k} denotes the CPU frequency of edge device D_k.

9.3.1.2 Upload Local Models

Following local training, each updated local model is encapsulated into a transaction, comprising a digital signature and the local model itself. The computation latency for generating the digital signature is given by:

$$T_{\text{up}}^{\text{cmp}} = \max_{D_k \in \mathcal{D}} \left(\frac{\rho}{f_{D_k}} \right). \tag{9.6}$$

Here, ρ represents the number of CPU cycles required for digital signature generation, and f_{D_k} is the CPU frequency of edge device D_k. We also define ϖ as the average size of one transaction, which correlates with the size of the AI model. The latency for transmitting data from an edge device to the primary edge server is thus expressed as:

$$T_{\text{up}}^{\text{com}} = \max_{D_k \in \mathcal{D}} \left\{ \frac{\varpi}{R_{D_k,B_p}^t} \right\}, \tag{9.7}$$

where $B_p \in \mathcal{B}$ denotes the primary edge server, and R_{D_k,B_p}^t is the transmission rate from edge device D_k to the primary edge server B_p in round t.

9.3.1.3 Global Model Aggregation

The transactions, which include local models, are sent to the primary edge server. This server then verifies the digital signatures of these transactions to confirm the identities of the senders. Subsequently, it executes a smart contract to aggregate the local models into a global model. This smart contract, composed of immutable and trackable lines of code, facilitates secure global model aggregation, for instance, using the multi-KRUM algorithm. Once the global model is generated, the primary edge server packages all the local model transactions and the global model transaction into a new block.

The maximum number of transactions that a block can contain is $K+1$, where K is the number of edge devices (i.e., the number of local models) plus one transaction for the global model. Therefore, the block size, excluding the relatively small size of the block header, is $S_B = (K+1)\varpi$. In this phase, the primary edge server's tasks include verifying the digital signatures of K transactions and executing the smart contract for secure global model aggregation, which requires a total of σ CPU cycles. Consequently, the total computational workload at the primary edge server is $\Delta_{\text{req},B_p} = K\rho + \sigma$. The computation delay at the primary edge server is thus given by:

$$T_{\text{agg}}^{\text{cmp}} = \frac{\Delta_{\text{req}, B_p}}{f_{B_p}}. \tag{9.8}$$

9.3.1.4 Pre-prepare

After the new block is created, the primary edge server broadcasts a pre-prepare message to all validator edge servers for validation. This message includes the primary edge server's digital signature and new block information. The validator edge servers, constituting all edge servers except the primary one, are tasked with verifying the primary edge server's creation of the new block and the authenticity of all transactions within it, ensuring the block's validity. Additionally, these validators are responsible for verifying the correctness of the global model. This typically involves recalculating the global model using the local models and comparing the result with that provided by the primary edge server.

The process progresses to the next step if the pre-prepare message is validated by at least $2f$ validator edge servers, where f represents the number of hypothetical malicious edge servers in a system with $M = 3f + 1$ servers, as per the PBFT protocol [20]. PBFT is designed to guarantee the correctness of actions with the consensus of $2/3$ of the edge servers. This approach, which does not require the validation of every single edge server, is a balance between efficiency and security, providing a more efficient but slightly less secure method compared to full validation.

The latency of this step can be formulated as

$$T_{\text{prep}}^{\text{com}} = \max_{B_m \in \mathcal{B}/B_p} \left\{ \frac{S_B}{R_{B_p, B_m}^t} \right\}. \tag{9.9}$$

The amount of computation at the validator edge servers is $\Delta_{\text{prep}, B_m} = \rho + (K + 1)\rho + \sigma$, where there is $B_m \neq B_p$. The computation latency in this step is

$$T_{\text{prep}}^{\text{cmp}} = \max_{B_m \in \mathcal{B}/B_p} \left\{ \frac{\Delta_{\text{prep}, B_m}}{f_{B_m}} \right\}. \tag{9.10}$$

9.3.1.5 Prepare

Upon accepting the new block from the primary edge server, each validator edge server broadcasts a prepare message to all other edge servers, including both primary and validator servers. The purpose is to validate the prepare message for consistency with the pre-prepare message.

The communication latency in this step involves broadcasting the prepare message to all other validator edge servers. It is calculated as:

$$T_{\text{pre}}^{\text{com}} = \max_{B_m \in \mathcal{B}/B_p, B_{m'} \in \mathcal{B}, B_m \neq B_{m'}} \left\{ \frac{S_M}{R^t_{B_m, B_{m'}}} \right\}, \tag{9.11}$$

where S_M is the size of the prepare message, which is smaller than S_B as it does not contain the new block. $R^t_{B_m, B_{m'}}$ represents the transmission rate between two different edge servers in the round t.

For computation, the primary edge server validates $2f$ digital signatures from other validator edge servers, requiring computational work of $\Delta_{\text{pre}, B_p} = 2f\rho$. Similarly, each validator edge server generates one digital signature for the prepare message and validates $2f$ digital signatures, leading to a total computational requirement of $\Delta_{\text{pre}, B_m} = \rho + 2f\rho$ for each validator server B_m (where $B_m \neq B_p$). The computation latency for this step is given by:

$$T_{\text{pre}}^{\text{cmp}} = \max_{B_m \in \mathcal{B}} \frac{\Delta_{\text{pre}, B_m}}{f_{B_m}}. \tag{9.12}$$

9.3.1.6 Commit

Once receiving $2f$ valid prepare messages from other validator edge servers are received, each edge server broadcasts a commit message to all other servers in the network. The communication latency for this broadcasting step is represented by

$$T_{\text{cmit}}^{\text{com}} = \max_{B_m \in \mathcal{B}/B_p, B_{m'} \in \mathcal{B}, B_m \neq, B_{m'}} \left\{ \frac{S_M}{R^t_{B_m, B_{m'}}} \right\}. \tag{9.13}$$

In this step, each edge server is required to generate one digital signature to construct their commit messages. Upon receiving these messages, they must also verify $2f$ digital signatures. Consequently, the total computational requirement for each edge server B_m is $\Delta_{\text{cmit}, B_m} = \rho + 2f\rho$, where ρ represents the number of CPU cycles needed for a single digital signature operation. The computation latency in this step can be expressed as

$$T_{\text{cmit}}^{\text{cmp}} = \max_{B_m \in \mathcal{B}} \left\{ \frac{\Delta_{\text{cmit}, B_m}}{f_{B_m}} \right\}. \tag{9.14}$$

9.3.1.7 Reply

Once $2f$ commit messages are received, the new block is confirmed as valid and appended to the blockchain. A reply message is then sent to the primary edge server.

The communication latency for transmitting this reply message is given by

$$T_{\text{rep}}^{\text{com}} = \max_{B_m \in \mathcal{B}/B_p} \left\{ \frac{S_M}{R_{B_m,B_p}^t} \right\}. \tag{9.15}$$

Regarding computation, each validator edge server generates one digital signature for the reply message, resulting in a computational requirement of $\Delta_{\text{rep},B_m} = \rho$ for each. The primary edge server, on receiving these messages, needs to verify $2f$ digital signatures, necessitating a computational workload of $\Delta_{\text{rep},B_p} = 2f\rho$. The computation latency for this step is thus

$$T_{\text{rep}}^{\text{cmp}} = \max_{B_m \in \mathcal{B}} \left\{ \frac{\Delta_{\text{rep},B_m}}{f_{B_m}} \right\}. \tag{9.16}$$

9.3.1.8 Download Global Model

After the steps of local model validation, global model aggregation, and the addition of the new block to the blockchain are completed, the global model is then transmitted to the edge devices to initiate the next training round. The communication latency for this transmission step is

$$T_{\text{down}}^{\text{com}} = \max_{D_k \in \mathcal{D}} \left\{ \frac{\varpi}{R_{B_p,D_k}^t} \right\}. \tag{9.17}$$

From the above analysis, the total latency in the t-th round consists of communication latency and computation latency, i.e.,

$$T(\boldsymbol{b}^t, \boldsymbol{p}^t) = T^{\text{com}} + T^{\text{cmp}}. \tag{9.18}$$

The communication time consumption can be calculated as

$$T^{\text{com}} = T_{\text{up}}^{\text{com}} + T_{\text{prep}}^{\text{com}} + T_{\text{pre}}^{\text{com}} + T_{\text{cmit}}^{\text{com}} + T_{\text{rep}}^{\text{com}} + T_{\text{down}}^{\text{com}}, \tag{9.19}$$

and the computation time consumption is

$$T^{\text{cmp}} = T_{\text{train}}^{\text{cmp}} + T_{\text{up}}^{\text{cmp}} + T_{\text{agg}}^{\text{cmp}} + T_{\text{prep}}^{\text{cmp}} + T_{\text{pre}}^{\text{cmp}} + T_{\text{cmit}}^{\text{cmp}} + T_{\text{rep}}^{\text{cmp}}. \tag{9.20}$$

9.3.2 *Problem Formulation and Analysis*

In this section, our goal is to minimize the long-term average latency of wireless B-FEEL by optimizing bandwidth allocation and transmit power allocation. We consider that the wireless B-FEEL training process is completed in the τ-th training round. The problem of minimizing the long-term average latency is formulated as follows

$$\underset{\{\boldsymbol{b}^t,\boldsymbol{p}^t\}_{t\in\mathcal{T}}}{\text{minimize}} \quad \frac{1}{\tau}\sum_{t=1}^{\tau} T(\boldsymbol{b}^t,\boldsymbol{p}^t)$$

$$\text{subject to} \quad \sum_{d\in\mathcal{D}\cup\mathcal{B}} b_d^t \le b^{\max}, b_d^t \ge 0, \forall t \in \mathcal{T}, \tag{9.21a}$$

$$\frac{1}{\tau}\sum_{t=1}^{\tau}\sum_{d\in\mathcal{B}\cup\mathcal{D}} p_d^t \le \bar{p}, p_d^t \ge 0, \tag{9.21b}$$

where $\mathcal{T} = \{1, 2, \ldots, \tau\}$ denotes the training round. In the above formulation, (9.21a) represents the maximum system bandwidth constraint and (9.21b) restricts the long-term average power.

Problem (9.21) presents a long-term resource allocation challenge due to the correlation in the long-term average power constraint. It cannot be transformed into multiple one-shot problems, and the optimization variables are interconnected in both the objective function and constraints. Conventional optimization-based algorithms face high computational complexity, which is unsuitable for time-sensitive B-FEEL systems. To address these challenges, we will introduce a DRL-based algorithm aimed at minimizing the long-term cumulative latency.

9.4 TD3 Based Resource Allocation Algorithm

In this section, we begin by formulating the optimization problem as an MDP. Subsequently, we introduce a TD3-based learning algorithm designed to efficiently solve the MDP problem.

9.4.1 MDP Formulation

To find optimal solutions within a stochastic environment, we have transformed problem (9.21) into an MDP, amenable to a DRL based approach. The MDP framework encompasses essential elements such as the observed state, action, and reward, which are outlined as follows.

- *State*: In each training phase, edge servers collect vital channel state information (CSI) and latency data from devices, crucial for resource management. The system state at time t, denoted as s^t, is defined by these cumulative latency and CSI metrics. Specifically, it includes channel gains between edge devices and the primary server ($\{h^t_{d,B_p}\}, \forall d \in \mathcal{D}$) with K entries, and between different validator edge servers ($\{h^t_{d_1,d_2}\}, \forall d_1, d_2 \in \mathcal{B}, d_1 \neq d_2$) having $M(M-1)$ entries. Thus, the state is expressed as

$$s^t = \left\{\sum_{i=0}^{t-1} T(\boldsymbol{b}^i, \boldsymbol{p}^i), \{h^t_{d,B_p}\}, \forall d \in \mathcal{D}, \right. \\ \left. \{h^t_{d_1,d_2}\}, \forall d_1, d_2 \in \mathcal{B}, d_1 \neq d_2 \right\}. \tag{9.22}$$

Consequently, the state space dimension is $K + M(M-1) + 1$. The DRL-based resource allocation algorithm, proposed for this setup, is adaptable to correlated wireless channel models. However, in uncorrelated channels like the Rayleigh channel, the reduced correlation between consecutive states might impact the accuracy of future reward predictions and, subsequently, the efficiency of resource allocation decisions, as suggested in the study by Wang et al. [28].

- *Action*: In this setup, the agent selects an optimal action based on the observed environmental states, guided by the policy π. The action comprises bandwidth allocation $\boldsymbol{b}^t$ and transmit power allocation $\boldsymbol{p}^t$, both in $\mathbb{R}^{M+K}$, for all edge servers and devices at the training round t. This is mathematically expressed as:

$$a^t = \left\{\boldsymbol{b}^t, \boldsymbol{p}^t\right\}.$$

As a result, the dimensionality of the action space is quantified as $2(M+K)$, encapsulating the variables of bandwidth and transmit power allocation for each participant in the network.

- *Reward*: Upon executing action a^t, the DRL agent receives a reward from the environment, which is then used to update policy π for maximizing future rewards. To reduce latency in the wireless B-FEEL training process, the reward at step t is inversely related to latency; a higher reward corresponds to lower latency. This is formulated as

$$r^t = \begin{cases} -T(\boldsymbol{b}^t, \boldsymbol{p}^t) & \text{when (9.21a), (9.21b) are satisfied,} \\ r_p & \text{otherwise,} \end{cases} \tag{9.23}$$

Here, when constraints (9.21a) and (9.21b) are fulfilled, the immediate reward is the negative of the latency in that round. If these constraints are not met, the immediate reward is set to r_p, a significantly small value acting as a penalty. This approach ensures that the actions chosen by the agent adhere to the constraints (9.21a) and (9.21b).

- *MDP Optimization Problem*:The action-value function in your scenario is defined as:

$$Q_{\pi_\phi}(s^t, a^t) = \mathbb{E}_{\pi_\phi}\left[\sum_{i=0}^{\infty} \gamma^i r^{t+i} | s^t, a^t\right], \tag{9.24}$$

where $\sum_{i=0}^{\infty} \gamma^i r^{t+i}$ represents the total discounted reward from the i-th training round, and $\gamma \in [0, 1]$ is the discount factor. This action-value function exhibits a "myopic" evaluation for $\gamma \to 0$ and a "far-sighted" evaluation for $\gamma \to 1$. It is used to assess the actions selected by a policy π_ϕ with parameters ϕ. Additionally, this function can be reformulated as the Bellman equation:

$$Q_{\pi_\phi}(s^t, a^t) = \mathbb{E}_{\pi_\phi}\left[r^t + \gamma Q_{\pi_\phi}(s^{(t+1)}, a^{(t+1)})|s^t, a^t\right]. \quad (9.25)$$

To solve the MDP problem, the optimal policy is derived by maximizing the action-value function

$$\pi_* = \arg\min_{\pi_\phi} Q_{\pi_\phi}(s^t, a^t). \quad (9.26)$$

To this end, the resource allocation problem is redefined as an MDP without pre-defined state transition probabilities $P(s^{t+1}|s^t, a^t)$. Consequently, model-free reinforcement learning based on value function approximation can be utilized to solve it, which bypasses the need for knowing transition probabilities to make decisions.

9.4.2 TD3 Algorithm

The TD3 is widely used to solve complex resource allocation problem in real-time. We thus adopt a TD3 based algorithm to solve the MDP problem of B-FEEL system. This implementation is detailed in Algorithm 13. TD3 uses an actor-critic architecture, integrating value-based and policy-based reinforcement learning methods, as depicted in Fig. 9.3. In this structure, the actor outputs actions in a continuous space derived from the observed state, following the policy $a^t = \pi_\phi(s^t)$. Meanwhile, the critic approximates the action-value function, producing Q values

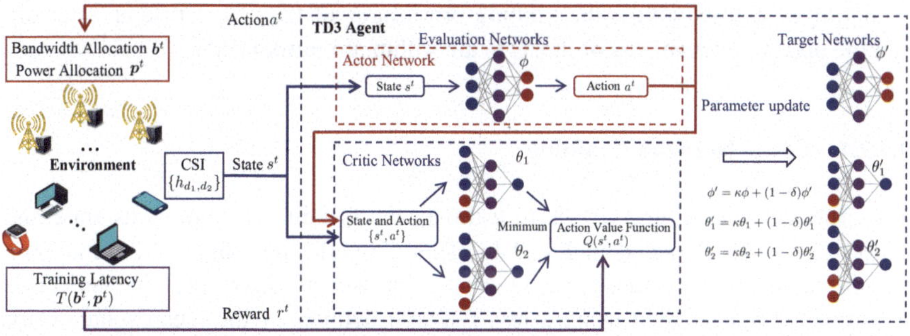

Fig. 9.3 Framework of TD3 algorithm

Algorithm 13: TD3-based bandwidth and transmit power allocation algorithm

1. **Ensure:** Actor network π_ϕ and critic networks $\{Q_{\theta_1}, Q_{\theta_2}\}$
2. Initialize critic networks $\{Q_{\theta_1}, Q_{\theta_2}\}$ and actor network π_ϕ with random parameters $\theta_1, \theta_2, \pi_\phi$.
3. Initialize target networks $\theta_1' \leftarrow \theta_1, \theta_2' \leftarrow \theta_2, \phi' \leftarrow \phi$.
4. Initialize replay buffer $\mathcal{R}$.
5. Initialize noise range c, noise variance σ_1^2, σ_2^2, update proportion δ, discount factor γ.
6. Explore E time steps with random policy, obtain the reward r^t and the next state s^{t+1}, and store the transition (s^t, a^t, r^t, s^{t+1}) of the E steps into $\mathcal{R}$.
7. Repeat until time step $\leq P$:
   - Select action with exploration noise: $a^t = \pi_\phi(s^t) + n_1$, where $n_1 \sim \mathcal{N}(0, \sigma_1^2)$.
   - Obtain bandwidth allocation $\boldsymbol{b}^t$ and transmit power allocation $\boldsymbol{p}^t$ from action a^t, and assign them to edge servers and device devices.
   - Edge devices and edge servers begin B-FEEL training process and get the latency in one round.
   - Calculate reward r^t, update state s^{t+1}, and store the transition (s^t, a^t, r^t, s^{t+1}) into $\mathcal{R}$.
   - Sample a mini-batch of transitions from $\mathcal{R}$.
   - Obtain target action $a'^{t+1} = \pi_{\phi'}(s^{t+1}) + n_2$, where $n_2 \sim \text{clip}(\mathcal{N}(0, \sigma_2^2), -c, c)$.
   - Calculate $y = r^t + \gamma \min_{i=1,2} Q_{\theta_i'}(s^{t+1}, a'^{t+1})$.
   - Update critics $\theta_i \leftarrow \arg\min_{\theta_i} \mathbb{E}\left[(y - Q_{\theta_i}(s^t, a^t))^2\right]$.
   - Update ϕ using the deterministic policy gradient algorithm.
   - If step mod $\vartheta = 0$:
     a. Update the target networks:
     b. $\theta_1' = \kappa\theta_1 + (1-\kappa)\theta_1'$
     c. $\theta_2' = \kappa\theta_2 + (1-\kappa)\theta_2'$
     d. $\phi' = \kappa\phi + (1-\kappa)\phi'$
   - Increment $t \leftarrow t + 1$.

based on the selected action and current state, namely $Q(s^t, a^t)$. Given the intricate mapping functions of the actor and critic, as well as the vectorized state and action, both components are realized through deep neural networks (DNNs). These DNNs consist of multiple fully-connected layers forming the feature extractor and the output layer. Subsequently, the training procedures of the actor-critic architecture and the update mechanisms of the TD3 algorithm are introduced.

9.4.2.1 Critic Network Training

The critic network's objective is to estimate the Q table in continuous state and action spaces, as depicted in Fig. 9.4a. This network approximates the mapping from state and action values to the Q value, expressed as $Q_\theta(s^t, a^t)$. The network's performance is evaluated using a loss function defined by the mean square error

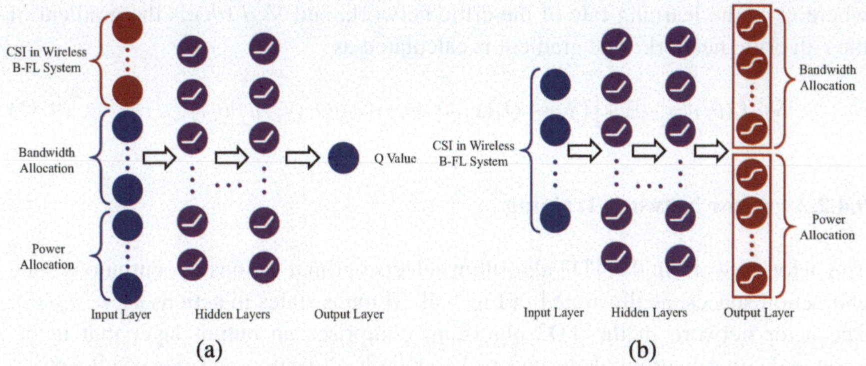

Fig. 9.4 Proposed structures of actor and critic networks in TD3 algorithm. (**a**) Structure of critic network. (**b**) Structure of actor network

(MSE) between the estimated and target values, formulated as

$$L(\theta) = \mathbb{E}\left[\left(y - Q_\theta(s^t, a^t)\right)^2\right], \tag{9.27}$$

where y is the target value and $Q_\theta(s^t, a^t)$ is the estimated value derived from the DNN parameter θ. Significantly, y is calculated using the target critic network

$$y = r^t + \gamma Q_{\theta'}(s^{t+1}, a'^{t+1}), \tag{9.28}$$

Here, θ' denotes the DNN parameters in the target network of the critic, and γ is the discount factor. The parameters θ' of the target critic network $Q_{\theta'}(s^t, a'^t)$ are periodically updated, and a'^{t+1} represents the action chosen for the subsequent state s^{t+1} by the target actor network $\pi_{\phi'}$. This structure enables a more stable and reliable training process for the critic network.

The update of the value function using the target value specified earlier can introduce an overestimate bias, which may affect the accurate update of the actor. To mitigate this bias issue, the TD3 algorithm employs two approximately independent target critics, $\{Q_{\theta_1}, Q_{\theta_2}\}$, for value function estimation. Here, Q_{θ_1} acts as a biased estimator, while Q_{θ_2} has a reduced overestimate bias. The minimum of these two estimators is utilized to update the value function

$$y = r^t + \gamma \min\left\{Q_{\theta'_1}(s^{t+1}, a'^{t+1}), Q_{\theta'_2}(s^{t+1}, a'^{t+1})\right\}. \tag{9.29}$$

In accordance with the loss function in (9.27), the update rules for the two critic networks are defined as:

$$\theta_i = \theta_i - \eta_c \nabla_{\theta_i} L(\theta_i), \forall i = 1, 2, \tag{9.30}$$

where η_c is the learning rate of the critic network, and $\nabla_{\theta_i} L(\theta_i)$ is the gradient of the i-th critic network. The gradient is calculated as

$$\nabla_{\theta_i} L(\theta_i) = -\mathbb{E}_{s^t}\left[\left(y_j - Q_i(s_j, a_j|\theta_i)\right)\nabla_{\theta_i} Q_i(s_j, a_j|\theta_i)\right]. \tag{9.31}$$

9.4.2.2 Actor Network Training

The actor network in the TD3 algorithm selects optimal actions in continuous state and action spaces, as illustrated in Fig. 9.4b. It maps states to actions $a^t = \pi_\phi(s^t)$. The actor network in the TD3 algorithm comprises an output layer that incorporates distinct neurons dedicated to bandwidth allocation and power allocation, respectively. For bandwidth allocation, a subset of neurons employs the softmax function, guaranteeing that their collective outputs sum to 1, thereby representing the percentage distribution of bandwidth. Meanwhile, for power allocation, another subset of neurons utilizes the Sigmoid function, ensuring that each output falls within the range of 0 to 1, indicative of the percentage of maximum transmit power allocated. The selection of the optimal action is geared towards maximizing the expected value $\mathbb{E}\left[Q_1(s^t, a^t|\theta_1)\right]$, employing the DNN parameters of the first critic network for updating the actor network. Accordingly, we update the actor via the deterministic policy gradient algorithm (DPG) as follows

$$\nabla_\phi J(\phi) = \mathbb{E}_{\pi_\phi}\nabla_a\left[Q_{\pi_\phi}(s^{t+1}, a^{t+1})|_{a=\pi_\phi(s^t)}\nabla_\phi\pi_\phi(s^t)\right], \tag{9.32}$$

where $\pi_\phi(s^t)$ is actor network with DNN parameter ϕ. Therefore, the update rule of actor network is

$$\phi = \phi - \eta_a\nabla_\phi J(\phi), \tag{9.33}$$

where η_a is learning rate of actor network.

9.4.2.3 Target Networks Updating

In the TD3 algorithm, both the two critic networks and the actor network have their respective target networks: the target critic networks and the target actor network. The target critic networks are responsible for generating target values that approximate the real Q values, while the target actor network is utilized for action selection. The critic and actor networks, also known as the online networks, undertake the crucial role of updating the DNN parameters at each step. In contrast, the target networks, which undergo updates less frequently, serve as a stable reference point for the real Q values and optimal actions. The DNN parameters of these target networks are updated based on the online networks, but in a delayed manner; specifically, they are only updated after the online networks have taken several steps. This delayed update ensures stability in the learning process. The general update rule for the target networks is as follows

$$\theta_1' = \kappa\theta_1 + (1-\kappa)\theta_1', \tag{9.34}$$

$$\theta_2' = \kappa\theta_2 + (1-\kappa)\theta_2', \tag{9.35}$$

$$\phi' = \kappa\phi + (1-\kappa)\phi', \tag{9.36}$$

where κ represents the update proportion, indicating the extent to which the DNN parameters of the online networks are integrated into the target networks. This method ensures gradual and stable updates to the target networks, contributing to the overall efficiency and effectiveness of the learning process.

9.5 Numerical Results and Analysis

9.5.1 Settings and Benchmarks

9.5.1.1 System Settings

We model a circular wireless network comprising $M = 4$ edge servers, catering to $K = 10$ edge devices. These servers and devices are uniformly spread within a circle that has a radius of 100 meters. We adhere to the LTE standard for the time-slot duration, setting T_0 at 10ms. Considering the low mobility characteristic of the B-FEEL system, the Doppler frequency f_d is fixed at 5Hz. Additionally, the path loss exponent for the network is assumed to be $\alpha = 2.5$. Edge servers operate at a maximum CPU frequency of 2.4 GHz, while edge devices function at up to 1 GHz. The network's maximum transmission power is capped at 24 dBm, and the overall bandwidth available is 100 MHz. To complete the setup, the noise power spectral density in the system is $N_0 = -174$ dBm/Hz, which aligns with typical values in wireless communication scenarios. These parameters are carefully chosen to simulate a realistic environment for assessing the B-FEEL system's performance.

9.5.1.2 TD3 Algorithm Settings

The actor network is composed of five hidden layers with 512, 1024, 2048, 1024, and 512 neurons, respectively, using ReLU activation. The output layer utilizes a softmax function to generate actions within constraints. The critic network consists of four hidden layers with 512, 1024, 512, and 512 neurons, employing ReLU activation, and a linear output layer for Q value estimation. The hyperparameters for TD3 are set as follows: a replay buffer size of 10^6, update proportion $\kappa = 5 \times 10^{-3}$, discount factor $\gamma = 0.99$, exploration steps $E = 512$, update frequency $\vartheta = 2$, and learning rates for both the actor and critic networks $\eta_a = \eta_c = 1 \times 10^{-4}$.

9.5.1.3 Benchmarks

To evaluate the effectiveness of our TD3-based resource allocation algorithm, we conducted comparisons with three other algorithms:

- *Random Allocation*: The bandwidth and transmit power are randomly allocated to edge servers and devices, following a uniform distribution.
- *Average Allocation*: In this approach, resources, including bandwidth and transmit power, are evenly distributed among all edge servers and edge devices, resulting in a uniform allocation for each.
- *Monte Carlo Algorithm*: This algorithm entails randomly generating and evaluating C allocation solutions, ultimately selecting the one that exhibits the lowest latency. The quality of the chosen solution becomes closer to the global optimum as C increases, and in our experiments, we set C to 10^6.

9.5.2 *Performance of Wireless B-FEEL*

We implemented the B-FEEL processes according to Algorithm 12. We introduced a factor of malicious behavior among edge devices, where these devices upload local models with random DNN parameters. These parameters are sampled from a normal distribution $\mathcal{N}(0, 1)$.

We conducted simulations with varying percentages of malicious edge devices, ranging from 0 to 100%, while training CNN models over 100 global epochs. The results, presented in Table 9.1, demonstrate a noticeable decrease in accuracy as the proportion of malicious edge devices increases. Notably, the proposed B-FEEL approach exhibits robustness against malicious behavior, with no significant impact on accuracy when malicious devices are below 50%. Figure 9.5 illustrates the test accuracy across training rounds, revealing that the proposed B-FEEL with 40% malicious devices and FEEL with 0% malicious devices perform equally well. This is attributed to the secure global aggregation process based on the multi-KRUM algorithm, which effectively eliminates the influence of malicious local models on the global model. Additionally, the number of aggregated local models has inconspicuous impact under these simulation settings.

Table 9.1 The accuracies of FEEL and B-FEEL with MINST dataset over selected percentages of malicious devices

Malicious %	0%	20%	40%	60%	100%
FEEL with FedAvg	97.92	92.38	86.42	11.35	9.74
B-FEEL with multi-KRUM	97.93	97.93	97.90	88.14	9.74

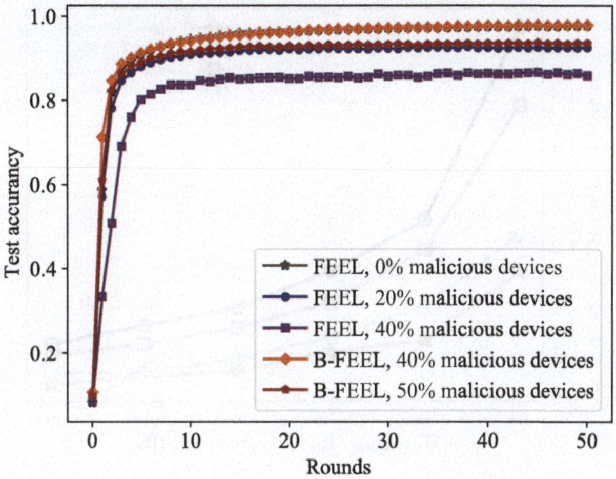

Fig. 9.5 Test accuracy of FEEL and B-FEEL for handwriting recognition

9.5.3 Performance Impact of Different System Settings

To demonstrate the effectiveness of our TD3-based resource allocation algorithm, we conduct a comparative analysis with three benchmark algorithms: random allocation, average allocation, and the Monte Carlo algorithm. We present three figures to visualize the performance of our proposed algorithm and its sensitivity to various system parameters, including maximum system bandwidth, maximum system transmit power, and the number of edge devices. To mitigate the influence of device location and channel state variability, our results are averaged across 500 system realizations.

Figure 9.6 illustrates the impact of the maximum system bandwidth, denoted as $b^{\max}$, on long-term latency during the B-FEEL training process. Firstly, it's evident that our proposed TD3-based algorithm yields similar results to the benchmark Monte Carlo algorithm and outperforms random allocation and average allocation schemes. Notably, the DRL-based algorithm slightly outperforms the Monte Carlo algorithm due to its computational efficiency and ability to choose a more optimal resource allocation scheme. Secondly, as $b^{\max}$ increases, the long-term latency significantly decreases.

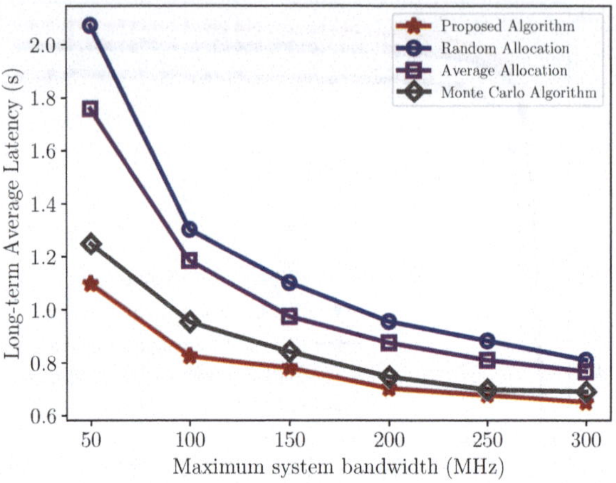

Fig. 9.6 Latency (s) vs. maximum system bandwidth (MHz)

9.6 Conclusion

In this chapter, we developed a B-FEEL architecture to ensure security and privacy, which utilizes secure global aggregation and blockchain to safeguard against potential attacks from malicious edge devices and servers. We integrated the PBFT consensus protocol into B-FEEL to achieve high effectiveness and low energy consumption, enhancing the trustworthiness of FEEL. To address the challenge of minimizing long-term average training latency, we formulated an optimization problem that jointly considers bandwidth and transmit power allocation. In addition, we derived a TD3-based algorithm for efficient resource allocation with low computational complexity. Finally, we conducted simulations to assess the learning performance of wireless B-FEEL and the efficiency of the DRL-based resource allocation algorithm.

References

1. J. So, B. Guler, and A. S. Avestimehr, "Byzantine-resilient secure federated learning," *IEEE J. Select. Areas Commun.*, vol. 39, pp. 2168–2181, Jul. 2021.
2. P. Blanchard, E. M. E. Mhamdi, R. Guerraoui, and J. Stainer, "Machine learning with adversaries: Byzantine tolerant gradient descent," in *Proc. Adv. Neural Inf. Process. Syst. (NeurIPS)*, pp. 118–128, Dec. 2017.
3. F. Guo, F. R. Yu, H. Zhang, H. Ji, M. Liu, and V. C. M. Leung, "Adaptive resource allocation in future wireless networks with blockchain and mobile edge computing," *IEEE Trans. Wireless Commun.*, vol. 19, pp. 1689–1703, Mar. 2020.
4. Z. Xiong, S. Feng, W. Wang, D. Niyato, P. Wang, and Z. Han, "Cloud/fog computing resource management and pricing for blockchain networks," *IEEE Internet Things J.*, vol. 6, pp. 4585–4600, Jun. 2019.

5. J. Kang, Z. Xiong, D. Niyato, D. Ye, D. I. Kim, and J. Zhao, "Toward secure blockchain-enabled internet of vehicles: Optimizing consensus management using reputation and contract theory," *IEEE Trans. Veh. Technol.*, vol. 68, pp. 2906–2920, Mar. 2019.
6. Z. Xiong, Y. Zhang, D. Niyato, P. Wang, and Z. Han, "When mobile blockchain meets edge computing," *IEEE Commun. Mag.*, vol. 56, pp. 33–39, Aug. 2018.
7. Y. Xiao, N. Zhang, W. Lou, and Y. T. Hou, "A survey of distributed consensus protocols for blockchain networks," *IEEE Commun. Surv. Tuts.*, vol. 22, pp. 1432–1465, Jan. 2020.
8. S. Warnat-Herresthal, H. Schultze, K. L. Shastry, S. Manamohan, S. Mukherjee, V. Garg, R. Sarveswara, K. Händler, P. Pickkers, N. A. Aziz, *et al.*, "Swarm learning for decentralized and confidential clinical machine learning," *Nature*, vol. 594, pp. 265–270, Jun. 2021.
9. D. C. Nguyen, M. Ding, Q.-V. Pham, P. N. Pathirana, L. B. Le, A. Seneviratne, J. Li, D. Niyato, and H. V. Poor, "Federated learning meets blockchain in edge computing: Opportunities and challenges," *IEEE Internet Things J.*, vol. 8, pp. 12806–12825, Aug. 2021.
10. Z. Wang and Q. Hu, "Blockchain-based federated learning: A comprehensive survey," *arXiv preprint arXiv:2110.02182*, Oct. 2021.
11. S. Zhou, H. Huang, W. Chen, P. Zhou, Z. Zheng, and S. Guo, "Pirate: A blockchain-based secure framework of distributed machine learning in 5g networks," *IEEE Network*, vol. 34, pp. 84–91, Nov. 2020.
12. K. B. Letaief, W. Chen, Y. Shi, J. Zhang, and Y.-J. A. Zhang, "The roadmap to 6g: Ai empowered wireless networks," *IEEE Commun. Mag.*, vol. 57, pp. 84–90, Aug. 2019.
13. Y. Shi, K. Yang, Z. Yang, and Y. Zhou, *Mobile Edge Artificial Intelligence: Opportunities and Challenges*. Elsevier, Aug. 2021.
14. L. Chang, Z. Zhang, P. Li, S. Xi, W. Guo, Y. Shen, Z. Xiong, J. Kang, D. Niyato, X. Qiao, *et al.*, "6g-enabled edge ai for metaverse: Challenges, methods, and future research directions," *J. Commun. Inf. Netw.*, vol. 7, pp. 107–121, Jun. 2022.
15. Z. Yang, Y. Zhou, Y. Wu, and Y. Shi, "Communication-efficient quantized sgd for learning polynomial neural network," in *Proc. IEEE Int. Perform. Comput. Commun. Conf. (IPCCC)*, pp. 1–6, Oct. 2021.
16. L. Li, L. Yang, X. Guo, Y. Shi, H. Wang, W. Chen, and K. B. Letaief, "Delay analysis of wireless federated learning based on saddle point approximation and large deviation theory," *IEEE J. Select. Areas Commun.*, vol. 39, pp. 3772–3789, Dec. 2021.
17. M. Chen, H. V. Poor, W. Saad, and S. Cui, "Convergence time optimization for federated learning over wireless networks," *IEEE Trans. Wireless Commun.*, vol. 20, pp. 2457–2471, Apr. 2021.
18. J. Ren, G. Yu, and G. Ding, "Accelerating dnn training in wireless federated edge learning systems," *IEEE J. Select. Areas Commun.*, vol. 39, pp. 219–232, Jan. 2021.
19. Z. Yang, M. Chen, W. Saad, C. S. Hong, M. Shikh-Bahaei, H. V. Poor, and S. Cui, "Delay minimization for federated learning over wireless communication networks," *arXiv preprint arXiv:2007.03462*, Jul. 2020.
20. M. Castro and B. Liskov, "Practical byzantine fault tolerance," in *Proc. USENIX Symp. Oper. Syst. Design Implement. (OSDI)*, vol. 99, pp. 173–186, Feb. 1999.
21. Z. Yang, Y. Shi, Y. Zhou, Z. Wang, and K. Yang, "Trustworthy federated learning via blockchain," *IEEE Internet of Things Journal*, vol. 10, no. 1, pp. 92–109, 2023.
22. X. Jiang, F. R. Yu, T. Song, and V. C. Leung, "Intelligent resource allocation for video analytics in blockchain-enabled internet of autonomous vehicles with edge computing," *IEEE Internet Things J.*, Sep. 2020. DOI: 10.1109/JIOT.2020.3026354.
23. T. Li, A. K. Sahu, A. Talwalkar, and V. Smith, "Federated learning: Challenges, methods, and future directions," *IEEE Signal Process. Mag.*, vol. 37, pp. 50–60, May. 2020.
24. C. Ma, J. Li, M. Ding, L. Shi, T. Wang, Z. Han, and H. V. Poor, "When federated learning meets blockchain: A new distributed learning paradigm," *arXiv preprint arXiv:2009.09338*, Sep. 2020.
25. R. Yang, F. R. Yu, P. Si, Z. Yang, and Y. Zhang, "Integrated blockchain and edge computing systems: A survey, some research issues and challenges," *IEEE Commun. Surv. Tuts.*, vol. 21, pp. 1508–1532, Jan. 2019.

26. M. Shayan, C. Fung, C. J. M. Yoon, and I. Beschastnikh, "Biscotti: A blockchain system for private and secure federated learning," *IEEE Trans. Parallel Distrib. Syst.*, vol. 32, pp. 1513–1525, Dec. 2020.
27. S. Huang, Y. Zhou, T. Wang, and Y. Shi, "Byzantine-resilient federated machine learning via over-the-air computation," in *Pro. IEEE Int. Conf. Commun. (ICC) Workshop*, pp. 1–6, IEEE, Jun. 2021.
28. Z. Wang, J. Zong, Y. Zhou, Y. Shi, and V. W. S. Wong, "Decentralized multi-agent power control in wireless networks with frequency reuse," *IEEE Trans. Commun.*, vol. 70, pp. 1666–1681, Mar. 2022.

The manufacturer's authorised representative in the EU is Springer
Nature Customer Service Centre GmbH, Europaplatz 3, 69115 Heidelberg,
Germany. If you have any concerns regarding our products, please
contact ProductSafety@springernature.com

Printed and bound by CPI Group (UK) Ltd, Croydon, CR0 4YY
30/09/2026
02225590-0003